Paulus Peronius Cato Hoek

The Voyage of H.M.S. Challenger

Report on the Pycnogonida

Paulus Peronius Cato Hoek

The Voyage of H.M.S. Challenger
Report on the Pycnogonida

ISBN/EAN: 9783741184420

Manufactured in Europe, USA, Canada, Australia, Japa

Cover: Foto ©Klaus-Uwe Gerhardt /pixelio.de

Manufactured and distributed by brebook publishing software
(www.brebook.com)

Paulus Peronius Cato Hoek

The Voyage of H.M.S. Challenger

THE

VOYAGE OF H.M.S. CHALLENGER.

ZOOLOGY.

REPORT on the Pycnogonida, dredged by H.M.S. Challenger during the years 1873–76. By Dr P. P. C. Hoek, Assistant at the Zootomical Laboratory of Leiden University.

The beautiful and rich collection of Pycnogonida formed during the expedition of the Challenger has been placed in my hands by Professor Sir Wyville Thomson, F.R.S., for description in the official report of the voyage.[1]

Our knowledge with regard to the Pycnogonids in general, their systematic arrangement, their geographical distribution, &c., is still very insufficient; and with respect to those of the greater depths of the ocean hardly anything is known.

The first attempt towards a monograph of the Pycnogonida is that of George Johnston.[2] His paper was published in 1837, and treats of the British species known up to that time. Though no special paper on Pycnogonids seems to have been published previous to Johnston's, yet there are several works of an older date, in which descriptions of species and genera belonging to this group occur, as well as discussions as to their place in the Zoological System. But as the descriptions are for the greater part very incomplete and the species therefore not to be recognised, these works are interesting only in so far as they show how much uncertainty has always been felt as to the place of the Pycnogonids among the Arthropoda.

Linnæus (1767)[3] brings the forms known to him under the genus *Phalangium*, in which also numerous land-spiders are placed, and which he ranges between *Hydrachna* and *Aranea* under his Insecta aptera.

[1] I wish to tender my sincere thanks to Professor Sir Wyville Thomson, F.R.S., &c., who liberally trusted to me—though a stranger—the drawing up of this report; at the same time to Mr John Murray who has kindly given me much valuable assistance.

[2] George Johnston.—An Attempt to ascertain the British Pycnogonidæ, in The Magazine of Zoology and Botany, conducted by W. Jardine, P. J. Selby, and G. Johnston, vol. i., 1837.

[3] Carolus Linnæus, Systema Naturæ, editio xii. rev., 1766.

Otho Fabricius (1780) [1] assigns to them the name *Pycnogonum* proposed by Brünnich, and places the *Cyamus ceti* with them in the same genus. He believes them to be most closely allied to Crustaceans.

J. C. Fabricius (1794) [2] places the two genera *Pycnogonum* and *Nymphon* along with *Pediculus, Acarus*, &c., in the eleventh class (the Antliata) of his entomological system.

Lamarck (1801) [3] gives the same genera (*Pycnogonum* and *Nymphon*) a place in the class of the Arachnida, order of the Palpistes, together with *Bdella, Acarus*, and *Hydrachna*.

Savigny (1816) [4] proposes to place the Pycnogonida among the Crustacea, an opinion afterwards embraced by Milne-Edwards (1834) [5] and Johnston (1837). According to Johnston, Savigny arrived at this conclusion by a very ingenious analysis of their organs. He pointed out that the proboscis of the *Pycnogonum* is a head, whereas the mandibles, palpi, and ovigerous organs are merely modifications of the legs, so that the Pycnogonida, like the Crustaceans, really have seven pair of legs, &c.

Johnston [6] himself, taking the assertions of Savigny as decisive, disagrees with those naturalists who object to the Pycnogonids being placed among the Crustaceans on account of the great simplicity of their anatomy. With Milne-Edwards he considers the Pycnogonids, although imperfect and even degraded, as formed on the same general plan as that of all the numerous other animals rightly placed in the class Crustacea.

There can be no doubt that Johnston's publication is one of the most important in the history of the knowledge of the group. Johnston gives a very clear description of the body of a Pycnogonid, fully discusses the systematic position of the order, proposes good characteristic marks for the genera, and enters into detailed descriptions of the species. The number of genera in his paper amounts to five (*Nymphon, Pallene, Orithyia, Phoxichilus, Pycnogonum*), each with one species, with the exception of the genus *Nymphon*, to which two species are assigned.

Of the authors who come after Johnston, Milne-Edwards is the first to be mentioned. In the third volume of his Histoire naturelle des crustacés (1840), he gives a very short description of the body of a Pycnogonid, and enumerates, but without paying special attention to the group, the species and genera known to him. Following Johnston as nearly as possible, he has the same five genera [7] and almost the same species. His descriptions are very insufficient ; his work derives importance only from the circumstance that he places—as I have already mentioned above—the Pycnogonids among the Crustaceans as a distinct order, viz., that of the Araneiformes.

[1] Othonis Fabricii Fauna Groenlandica, Hafniæ et Lipsiæ, 1780.
[2] Joh. Christ. Fabricii Entomologia Systematica emendata et aucta, tom. iv., 1794.
[3] J. B. Lamarck.—Système des animaux sans vertèbres, à Paris, an. ix., 1801.
[4] J. C. Savigny.—Mémoires sur les animaux sans vertèbres, première partie, 1816.
[5] H. Milne-Edwards.—Histoire naturelle des Crustacés, tom. i.-iii., 1834-40.
[6] In this introduction only the most important authors are mentioned ; a much fuller list is given by Johnston in his An Attempt, &c., and by Milne-Edwards, *loc. cit.*
[7] The name *Orithyia* of Johnston " étant déjà employé pour un autre genre de Crustacé," is changed by Milne-Edwards into *Phoxichilidium* (*l.c.*, p. 535).

The species of the English coast found (1842–44) a new monographer in Goodsir,[1] who in three consecutive papers enumerates a large number of species new to the fauna of the British Isles and to science in general. Two new genera (*Pephredo* and *Pasithoe*) are proposed by him, but owing to the want of detail Mr Goodsir's papers are of little value, for it is absolutely impossible to recognise either his new genera or his new species from such descriptions as he gives.

Of as little value is the list given by Hodge (1864),[2] in which all Goodsir's species are found, in addition to some new *Ammotheas* and species of his new genus *Achelia*. Since Hodge's list—though occasionally in English periodicals short descriptions of new species have been published—no special paper on the Pycnogonids of the English coast has appeared.

Those of the Norwegian coast found a very able describer in Kröyer (1845),[3] who gives very clear diagnoses of the genera and species. As a new genus he proposes *Zetes*, and the total number of species described by him is twelve. These descriptions were published without illustrations; but illustrations to the text may be found in Quoy and Gaimard's Voyages en Scandinavie, Laponie, &c., Zoologie, Crustacés, pl. xxxix. (1840).

For the Pycnogonids of Northern Europe and the coasts of the Arctic Ocean, besides Kröyer, the following authors must be mentioned :—Otho Fabricius[4] for the coast of Greenland, as mentioned above. Sabine[5] (1824) describes two *Nymphons* (*N. grossipes* and *N. hirsutum*) and a species of *Phoxichilus* (*P. proboscideus*—a true *Colossendeis*, Jarzynsky), found on the shores of the North Georgian Islands. Bell (1855),[6] in Belcher's Last of the Arctic Voyages, gives descriptions and drawings of two new species of *Nymphon* (*N. hirtipes* and *N. robustum*) common in higher northern latitudes. Jarzynsky (1870)[7] enumerates the species of Russian Lapland and the White Sea. A new genus (*Colossendeis*) is proposed by him. Buchholz (1874),[8] in the narrative of the second German North Polar Expedition, enumerates three species of *Nymphon*, but none of these are new.

---

[1] Harry D. S. Goodsir.—Edinburgh New Philosophical Journal, vol. xxxii., 1842 ; *ibid.*, vol. xxxiii., 1842; On the Specific and Generic Characters of the Araneiform Crustacea, Annals and Mag. of Nat. Hist., vol. xiv., 1844.

[2] George Hodge.—List of the British Pycnogonoidea, with descriptions of several new species, Ann. and Mag. of Nat. Hist., vol. xiii., 3d series, 1864.

[3] Henrik Kröyer.—Bidrag til Kundskab om Pyknogoniderne eller Sospindlerne, Natur-historisk Tidskrift, Ny Raekke, i., 1845.

[4] *Loc. cit.*

[5] A Supplement to the Appendix of Captain Parry's Voyage for the Discovery of a North-West Passage in the years 1819-20, containing the Zoological and Botanical Notices, London, 1824 ; Marine Invertebrate Animals, by Captain Edward Sabine.

[6] Thomas Bell.—Account of the Crustacea of the Last of the Arctic Voyages in Search of Sir John Franklin, under the command of Captain Sir E. Belcher, C.B., &c., in two volumes, vol. ii., 1855.

[7] Th. Jarzynsky.—Promissus catalogus Pycnogonidarum inventarum in mari glaciali ad oras Lapponiæ rossicæ et in Mari albo, anno 1869 et 1870, Annales de la Soc. des Natur. de St Petersb., 1870.

[8] R. Buchholz.—Crustaceen der Zweiten Deutschen Nordpolarfahrt, Anhang: Pycnogonida, Die Zweit Deutsche Nordpolarfahrt, ii. 396, 1874.

Heller (1875) [1] proposes two new species of the same genus gathered during the Austrian North Polar Expedition; both are identical with species described before under other names.

In 1877 and again in 1879 G. O. Sars [2] published lists of the Pycnogonids gathered during dredging cruises in the northern part of the North Atlantic, on the coast of Norway, &c.   There are in all four new species of *Nymphon* (*N. megalops*, *N. macronyx*, *N. serratum* and *N. pallenoides*), a new genus, *Ascorhynchus*, with the species *Ascorhynchus abyssi*, a new species of *Colossendeis* (*C. angusta*), and a new *Pallene*, *P. malleolata*.

Miers (1877) [3] treats of the Pycnogonids collected during the last English Arctic Expedition.   He gives two species, neither of which is new, and describes a variety of *Nymphon hirtum*.

In regard to the coast of Germany and the Netherlands not a single species has been recorded which is not found on the English coast.   Occasionally enumerations of species have been published by Frey and Leuckart, [4] and Böhm. [5]   In a paper I published myself (1877) [6] I described the four genera, species of which are found on the Dutch coast.

The Pycnogonids of the French coast have been studied by Quatrefages (1844), [7] Claparède (1863), [8] Hesse (1867-74), [9] and Grube (1868-72). [10]   Their studies resulted in the proposal of a new species of *Ammothea* (*A. pycnogonoides*, Quatr.), of a new *Phoxichilidium* (?) (*P. cheliferum*, Claparède), a new species of *Phoxichilus* (*P. lævis*, Grube), and two new genera (?) *Oiceobathes*, Hesse, and (?) *Oomerus*, Hesse, both very insufficiently described.   The Pycnogonids found on the coasts of France, the British Isles, Germany, &c., are not yet sufficiently well known to allow of their geographical distribution being discussed.

About the species of the Mediterranean very little is known.   Philippi (1843) [11] and

[1] Camil Heller.—Die Crustaceen, Pycnogoniden, und Tunicaten der K. K. Oester. Ungar. Nordpol. Expedition, Denkschriften der Mathematisch-Naturwiss. Classe der K. Akad. der Wissensch., Bd. xxxv., Wien, 1875.

[2] G. O. Sars.—Prodromus descriptionis Crustaceorum et Pycnogonidarum, quae in expeditione Norvegica, anno 1876, observavit, Arch. f. Math. og Naturvid., ii., 1877 ; Crustacea et Pycnogonida nova, quae in itinere 2do et 3tio expeditionis Norvegicæ, anno 1877 et 1878, collecta (Prodromus descriptionis), *ibid.*, iv., 1879.

[3] Edward J. Miers.—Report on the Crustacea collected by the Naturalists of the Arctic Expedition in 1875-76 Annals and Mag. of Nat. Hist., fourth series, vol. xx., 1877.

[4] Frey und Leuckart.—Beiträge zur Kenntniss wirbelloser Thiere, 1847.

[5] R. Böhm.—Ueber die Pycnogoniden des Königl. Zool. Museums zu Berlin, Monatsber. der Königl., Akad. der Wiss. 1879.

[6] P. P. C. Hoek.—Ueber Pycnogoniden, Niederl. Archiv. f. Zoologie, iii., 1877.

[7] A. de Quatrefages.—Mémoire sur l'organisation des Pycnogonides, Ann. d. Sc. Natur., 3me Série, Zoologie, tom. iv., 1845.

[8] A. René Edouard Claparède.—Beobachtungen über Anatomie und Entwickelungsgeschichte wirbelloser Thiere an der Küste von Normandie angestellt, 1863.

[9] Hesse.—Annales des Sciences naturelles, 5ième Série, vii., 1867 ; *ibid.* 5ième Série, xx., 1874.

[10] Edward Grube.—Mittheilungen über St Malo und Roscoff und die dortige Meeres besonders die Anneliden-fauna 1869 ; Mittheilungen über St Vaast la Hougue, und seine Meeres, besonders seine Anneliden-fauna, Verhandl. der Schlesischen Gesellsch. f. vaterl. Cultur, 1869.

[11] A. Phillipi.—Ueber die Neapolitanischen Pycnogoniden, Arch. f. Naturgesch, ix., 1843.

Costa (1838–61)[1] published short notes on the Pycnogonids found there. Philippi proposed a new genus (*Endeis*), which is perhaps identical with *Pasithoe*, Goodsir; and a second genus (*Pariboea*), with the species *Pariboea spinipalpis*. Costa introduces (1838) the genus *Phanodemus*, in all probability identical with *Pephredo*, Goodsir; in his Microdoride mediterranea (1861) he proposes three new genera : *Rhynchothorax*, *Platychelus*, and *Aleynous*. From the Gulf of Naples Costa knows in all seven species, whereas the total number of species of the Mediterranean found in Philippi's paper is only four. A monograph on the Pycnogonids of the Mediterranean, and especially of the Gulf of Naples, will very probably soon appear ; it will form the second part of the Studi e Ricerche di Cavanna (1877),[2] and will also be published by Dohrn, as announced in his Neue Untersuchungen (1878).[3]

Of all Pycnogonida, those found on the west coast of North America are best known. Careful attention was paid to them by Stimpson (1852),[4] Verrill, Smith (1874),[5] but especially by Wilson (1878–80),[6] who in his Pycnogonida of New England, enumerates fourteen species belonging to nine genera, two of which (*Pseudopallene* and *Anoplodactylus*) are new to science. Though I do not believe that these new genera after a careful examination will hold good, and though I think it a pity that Wilson in his researches has not taken advantage of recent investigations (especially those of Cavanna), yet there can be no doubt, I believe, that his paper is one of the best descriptive publications after those of Johnston and Kröyer.

For the other countries of our globe, a very brief enumeration may suffice. As far as I have been able to ascertain, by far the greater number of the species described are littoral ; from the open ocean very few species are recorded. Two species described by White (1847),[7] inhabiting the South Sea, are exceptions. White describes them as species of *Nymphon*, whereas I believe that they ought to be considered as *Phoxichilidiums*. From the open ocean are also those species (one of *Nymphon*, another of *Phoxichilidium*) mentioned by Grube (1869)[8] as occurring in the China Sea. Grube's descriptions as well as those of White are extremely incomplete.

Wood-Mason (1873)[9] described a species of a genus which he believed to be new,

[1] O. G. Costa.—Fauna del Regno di Napoli, Crostacei et Aracnedi, Napoli, 1838 ; Microdoride mediterranea, tomo primo, Napoli, 1861.

[2] G. Cavanna.—Studi e Ricerche sui Pienogonidi, parte prima (Publicazioni del R. Istituto di Studi superiori pratici et di perfezionamento in Firenze, Sezione di Scienze fisiche e naturali), Firenze, 1877.

[3] A. Dohrn.—Neue Untersuchungen über Pycnogoniden, Mittheil. a. d. Zoologischen Station zu Neapel, i., 1879.

[4] William Stimpson.—Synopsis of the Marine Invertebrata of Grand Manan, Smithsonian Contributions to Knowledge, January 1853.

[5] Smith in Report on the Invertebrata of Vineyard Sound. In Part I. of the Report on the Condition of the Sea-Fisheries of the South Coast of New England, 1873.

[6] E. B. Wilson.—Descriptions of Two New Genera of Pycnogonida, American Journal of Science and Arts, vol. xv., 1878 ; Synopsis of the Pycnogonida of New England, Transactions of the Connecticut Academy, vol. v., 1880.

[7] Adam White.—Descriptions of New or Little-Known Crustacea in the Collection at the British Museum, Proceedings of the Zoological Society of London, part 15, 1847.

[8] E. Grube in Jahresbericht der Schlesischen Ges. für vaterländische Cultur, Breslau, 1869.

[9] James Wood-Mason.—On *Rhopalorhynchus kröyeri*, a new Genus and Species of Pycnogonida, with plate xiii., Journal of the Asiatic Society of Bengal, part 2, 1873.

and called *Rhopalorhynchus*. There can be no doubt that this is the same as the genus formerly (1870) described by Jarzynsky [1] as *Colossendeis*. Wood-Mason's species is an inhabitant of Port Blair, Andaman Islands.

Miers (1875 and 1879) [2] mentions two species of *Nymphon*, and one of a new genus, which he calls *Tanystylum*, and which is nearly allied to *Achelia*. These species were collected at Kerguelen Island during the visit of the English and American Transit of Venus expeditions to that Island. Böhm (1879) [3] has made a very careful study of the Pycnogonids of the Royal Zoological Museum at Berlin. He describes two species of *Nymphon* and one of *Achelia*, as collected at Kerguelen; one species of *Nymphon* collected south of the Cape of Good Hope, one *Pallene* (?) taken in the Straits of Magellan, another *Pallene* from Mozambique, a *Phoxichilidium* and a *Phoxichilus* collected in the neighbourhood of Singapore; finally, besides some species from Northern Europe, three species found near Enosima (Japan); one species of a new genus, which Böhm calls *Lecithorhynchus*, one *Ascorhynchus* (*Gnamptorhynchus*, Böhm), and one species of *Pallene*. Slater (1879) [4] published a short paper on a new genus of Pycnogonids (*Parazetes*) found in Japan, and described in the same paper a variety of *Pycnogonum litorale* from the same country.

In the Boston Journal of Natural History, Eights (1836 ?) mentions the genus *Decalopoda*, but I have not been able to ascertain whether this is a good genus, nor where it has been found. [5] A species of *Pasithoe* described by Dr Gould [6] is, according to Wilson (*loc.cit.* p. 2), "indeterminable." To Mr Wilson's paper I am also indebted for the mention of a species of Pycnogonid found on the coast of Chili: it seems to be a species of *Pycnogonum*. [7]

In this enumeration the reader will not find a complete list of the descriptive literature of Pycnogonida, but all the more important publications, together with the greater number of the minor papers on our group are mentioned. With a few exceptions the zoological publications about Pycnogonids are very superficial, and this I believe is owing partly to the circumstance that many authors who have had no opportunity of comparing large collections of different forms have published descriptions of species and even of genera from the examination of such species only. To describe new species, however, ought

---

[1] Th. Jarzynsky, *loc. cit.*

[2] E. J. Miers.—Descriptions of new species of Crustacea collected at Kerguelen's Island, by the Rev. A. E. Eaton. Annals and Magazine of Natural History, fourth series, vol. xvi., 1875 ; Crustacea of Kerguelen Island, Philosophical Transactions, London, vol. clxviii. ; extra volume, pp. 200–214, 1879, pl. xi.

[3] R. Böhm, *loc. cit.* ; the same in Sitzungsberichte der Gesellschaft naturforschender Freunde in Berlin, 1879, pp. 53 and 140.

[4] Henry H. Slater.—On a new genus of Pycnogon (*Parazetes*) and a variety of *Pycnogonum littorale* from Japan Ann. and Magaz. of Nat. History, 5th series, vol. iii., 1879.

[5] Boston Journal of Natural History, i. 204, t. 7. (See Cuvier's Animal Kingdom, London, Wm. S. Orr & Co., 1840, p. 468.)

[6] Proc. Boston Society Nat. Hist., vol. i. p. 92.

[7] Gay.—Historia fisica y politica de Chile, Zoologia, p. 308, pl. iv. fig. 8, 1854.

not to be the work of one who begins to study a group, as is often the case, but can only be done properly after laborious and continuous research. Moreover, the study of the literature is enormously encumbered by the circumstance, that descriptions of single species often lie buried in obscure periodicals. This circumstance, I hope, will be considered, when my report is found to be far from complete.

The collection of Pycnogonida brought home by the Challenger and placed in my hands numbers about 120 specimens. They were in an excellent state of preservation, and to facilitate the work, the bottles of spirit in which they were put, were furnished with labels indicating the station, latitude, longitude, bottom temperature, and the nature of the sea-bed where they were dredged. Some of the specimens were not obtained from any of the 361 dredging stations, but were collected on the shore (near Cape Town, e.g.), or dredged in shallow water (Bahia, Kerguelen). Over a course of 68,890 miles the dredge was let down at 361 stations, and Pycnogonids were procured on only twenty-six occasions. The 120 specimens of Pycnogonida brought home belong to thirty-six species, and thirty-three of these I have been obliged to consider and describe as new to science. The greatest depth where a Pycnogonid was found was 2650 fathoms; the greatest depth dredged during the cruise was 4575 fathoms.

In the following list I have given the range in depth at which species of Pycnogonida were found by the Challenger, and also recently during the cruise of the "Knight Errant":—

Shore,	.	.	*Discoarachne brevipes*, Hoek.
,,	.	.	*Hannonia typica*, Hoek.
7 to 20 fathoms,	.	.	*Phoxichilidium fluminense*, Kröyer.
,, ,, ,,		.	*Phoxichilidium insigne*, Hoek.
10 to 120 ,,		.	*Nymphon brachyrhynchus*, Hoek.
25 to 120 ,,		.	*Nymphon brevicaudatum*, Miers.
25 . ,,		.	*Nymphon fuscum*, Hoek.
38 . ,,		.	*Ascorhynchus minutus*, Hoek.
,, . ,,		.	*Pallene languida*, Hoek.
38 to 40 ,,		.	*Pallene laevis*, Hoek.
38 to 120 ,,		.	*Pallene australiensis*, Hoek.
45, 55, 175 ,,		.	*Phoxichilidium patagonicum*, Hoek.
53 . ,,		.	*Pycnogonum litorale*, Ström.
55, 70, 120 ,,		.	*Colossendeis megalonyx*, Hoek.
83 . ,,		.	*Nymphon brevicollum*, Hoek.
83 to 540 ,,		.	*Nymphon grossipes*, Oth. Fabr., sp.
120 . ,,		.	*Colossendeis robusta*, Hoek.
150 . ,,		.	*Ascorhynchus orthorhynchus*, Hoek.
375 to 540 ,,		.	*Nymphon robustum*, Bell.
400 to 1600 ,,		.	*Colossendeis leptorhynchus*, Hoek.
515, 530, 540 ,,		.	*Nymphon strömii*, Kröyer.
540 . ,,		.	*Nymphon macronyx*, G. O. Sars.
,, . ,,		.	*Colossendeis proboscidea*, Sab., sp.
600 . ,,		.	*Phoxichilidium patagonicum*, var. *elegans*, Hoek.

700	fathoms,	.	.	*Oorhynchus aucklandiæ*, Hoek.
825	.	,,	.	*Nymphon perlucidum*, Hoek.
1100	.	,,	.	*Nymphon longicoxa*, Hoek.
,,	.	,,	.	*Nymphon compactum*, Hoek.
1250	.	,,	.	*Colossendeis minuta*, Hoek.
1375	.	,,	.	*Ascorhynchus glaber*, Hoek.
1375 to 1600	,,	.	.	*Nymphon hamatum*, Hoek.
,,	,,	,,	.	*Colossendeis gigas*, Hoek.
,,	,,	,,	.	*Colossendeis gracilis*, Hoek.
1600 to 1950	,,	.	.	*Phoxichilidium pilosum*, Hoek.
1675	.	,,	.	*Nymphon meridionale*, Hoek.
,,	.	,,	.	*Phoxichilidium oscitans*, Hoek.
1875	.	,,	.	*Phoxichilidium mollissimum*, Hoek.
2160	.	,,	.	*Nymphon procerum*, Hoek.
2225	.	,,	.	*Nymphon longicollum*, Hoek.
,,	.	,,	.	*Colossendeis media*, Hoek.
2650	.	,,	.	*Colossendeis brevipes*, Hoek.

The number of times at which Pycnogonida were dredged at certain depths is shown in the following table :—

99 dredgings in depths of from			1 to 500 fathoms,	.	.	26 times.
30	,,	,,	501 to 1000 ,,	.	.	3 ,,
47	,,	,,	1001 to 1500 ,,	.	.	3 ,,
47	,,	,,	1501 to 2000 ,,	.	.	4 ,,
93	,,	,,	2001 to 2500 ,,	.	.	2 ,,
83	,,	,,	2501 to 3000 ,,	.	.	Once (at 2650 fathoms).
11	,,	,,	3001 to 4575 ,,	.	.	None.

It thus becomes apparent that what Davidson has shown for the Brachiopoda, holds also in the case of the Pycnogonida, that they are very seldom found in depths exceeding 500 fathoms ; out of about 100 dredgings in depths of from 1 to 500 fathoms, Pycnogonids were brought up twenty-six times, while in depths varying from 501 to 3000, they were obtained only thirteen times out of 300 dredgings.

The following statement shows the range in depth at which the genera of Pycnogonida hitherto known have been found. The total number is twenty-seven genera, of which eleven are true littoral forms. Of the sixteen remaining genera there are five of which I am quite uncertain as to the depth at which they are found, and four for which the depth does not exceed 50 fathoms. Then there are two (*Pallene* and *Pycnogonum*), which, as a rule, inhabit depths not exceeding 120 fathoms, but which in a single case were found at depths almost reaching 500 fathoms (*Pallene malleolata*, G. O. Sars, at a depth varying between 191 and 459 fathoms, and *Pycnogonum litorale*, dredged by Smith and Harger, at a depth of 430 fathoms). Hence there remain only five genera of Pycnogonida, species of which may truly be called deep-sea inhabitants; they are the genera *Nymphon*, *Ascorhynchus*, *Oorhynchus*, *Colossendeis*, and *Phoxichilidium*.

LIST OF THE GENERA OF PYCNOGONIDA HITHERTO KNOWN.

Name of the Genus.	Number of Species Described.	Depth in Fathoms at which they have been found.	Geographical Distribution.
Nymphon, Fabr.,	38	Shore to 2225.	Mundane—Pacific Ocean excepted.
Ammothea, Leach,	5	Shore to 5.	American and European Coasts of the North Atlantic.
Böhmia, Hoek,	1	(?)	(?)
Phanodemus, Costa,	3	Shore.	Coast of Italy.
Rhynchothorax, Cos.,	1	(?)	North Coast of Africa.
Pephredo, Goodsir,	1	Shore.	Coast of England.
Platychelus, Cos.,	1	(?)	Coast of Sardinia.
Oiceolathes, Hesse,	1	Shore.	Coast of France.
Ascorhynchus, G. O. Sars,	5	38 to 1539	North Atlantic, Indian Ocean, South Coast of Australia, North of New Guinea, Coast of Japan.
Zetes, Kröyer,	1	Shore.	Coast of Greenland.
Parazetes, Slater,	1	(?)	Japan.
Pariboea, Philippi,	1	Shore.	Coast of Italy.
Achelia, Hodge,	4	Shore to 35.	American and European Coasts of the North Atlantic, Coasts of the Mediterranean, Kerguelen.
Alcinous, Cos.,	2	(?)	Coast of Italy.
Tanystylum, Miers,	2	5 to 7	Kerguelen, East Coast of North America.
Lecithorhynchus, Böhm,	2	3 to 4	Japan.
Oorhynchus, Hoek,	1	700	Auckland.
Colossendeis, Jarzynsky,	12	55 to 2650	Mundane—Pacific Ocean excepted.
Pasithoe, Goodsir,	1	Shore.	Coast of England.
Endeis, Philippi,	2	Shore.	Coast of Italy.
Discoarachne, Hoek,	1	Shore.	Coast of Cape Colony.
Pallene, Johnston,	16	Shore to 459.[1]	Coast of Northern Europe, Greenland Sea, Coast of Greenland, East Coast of North America, Coast of Mozambique, off Australia, China Sea, Coast of Japan.
Phoxichilidium, M.-Edwards,	15	Shore to 1950.	Coast of Northern Europe, Greenland, North America, North Atlantic, Coast of Brazil, Patagonia, South Atlantic, Indian Ocean, Coast of Lower Siam, off Japan.
Oomerus, Hesse,	1	Shore.	Coast of France.
Hannonia, Hoek,	1	Shore.	Coast of Cape Colony.
Phoxichilus, Latr.,	4	Shore.	Coasts of Northern Europe, Mediterranean Coast, Coast of Lower Siam.
Pycnogonum, Brünnich,	2	Shore to 430.[2]	Coasts of Northern Europe, East Coast of North America, Coast of Chili, Coast of the Mediterranean, Coast of Australia.

[1] Teste G. O. Sars.   [2] Smith and Harger, teste Wilson.

When comparing the bathymetrical range of the different genera with their geographical distribution, it is easily remarked that it is the genera most widely spread over the bottom of the sea which are capable of existing at the greatest variety of depth. This is, for instance, the case with *Nymphon*, *Colossendeis*, and *Phoxichilidium*. Some species of *Nymphon* are found at low-water mark, others inhabiting shallow water in the immediate neighbourhood of the coast are dredged at a depth of under 100 fathoms; others again are never found at a depth exceeding 800 fathoms, and finally, there are some which are true deep-sea species. Some species of *Colossendeis* were dredged at a depth of under 100 fathoms, other species inhabit the ocean at a depth not exceeding 800 fathoms and others were dredged at depths varying from 800 to 2800 fathoms. The genus *Phoxichilidium* shows almost the same bathymetrical range as *Nymphon*. Now the geographical range of these three genera is, as far as I could ascertain from the facts at my disposal, nearly the same; this distribution is mundane. With the exception of the Pacific Ocean, from which as yet not a single species of Pycnogonid has been obtained, representatives of these three genera are found almost in every sea.

Of the genus *Ascorhynchus* only five species are known as yet. They were collected at depths varying from 38 to 1539 fathoms, and at widely distant places, viz., in the Greenland Sea, between the Cape of Good Hope and Kerguelen Islands, off Australia, to the north of New Guinea, and off Japan; and as the different species of this genus form a very natural group, it is, I think, very probable, that later investigations will show also for interjacent places the occurrence of forms belonging to this same group. *Oorhynchus* is as yet the only genus which seems to inhabit depths exceeding 800 fathoms exclusively; but as only a single specimen of the one species known of this genus has been collected, I do not think it expedient to pay much attention to this fact.

Hence, with regard to the bathymetrical range, a close study of the material brought home by the Challenger, added to what was previously known on the subject, has shown :—

(1) That those genera which range most widely geographically are also those which range most widely in depth; and (2), that there does not seem to exist a single true deep-sea genus of Pycnogonids.[1]

As for the influence of the increasing depth on the form and the structure of our animals, this is by no means easily traced. As far as the structure of the integument and of the eyes is concerned, I will treat the question at some length when speaking of their structure. As a rule the true deep-sea species are more slender, the legs very long and brittle, and the surface of the body smooth, whereas the true shore-inhabitants are much more concentrated, have shorter legs, and are often densely covered with

[1] From the study of deep-sea forms in general, Mr Moseley and others came to the conclusion that these animals have a world-wide range. Of this the Pycnogonids give a fair instance, I believe (Moseley in Nature, April 8, 1880, p. 546).

hair. However, these rules admit of a great many exceptions. Thus the most common shallow-water species of the English, French, and Dutch coasts is *Nymphon gracile*, Leach, an exceedingly slender animal with very long legs, and moreover almost smooth. *Colossendeis proboscidea*, Sab., sp., is a blind species occurring only at a considerable depth ; yet it has a highly concentrated body with short legs. Two species of *Phoxichilidium*, for which I have proposed the names *Phoxichilidium pilosum* and *Phoxichilidium mollissimum*, are true inhabitants of the depths of the ocean ; yet they are not smooth at all, but covered by a very hairy integument. The case of *Phoxichilidium patagonicum* and its variety *elegans*, which I describe hereafter, must probably be considered as a trifling instance of the effect of depth on the slenderness of the body.

The scientific and trustworthy material at our disposal is by no means sufficient to enable us to discuss thoroughly the question of the geographical distribution of Pycnogonids. Judging from what is known of the European and North American coasts, it is most probable that on all coasts, and everywhere in shallow water in the neighbourhood of the shore, forms of Pycnogonids will be found occurring ; and as I think it improbable that any true shore-inhabitant will be found which shows a very wide range, it is also highly probable that the number of rittoral forms at present known is very small in comparison with the number really existing.

The distribution of those Pycnogonids which are not to be considered as shore-inhabitants, but which have never been dredged yet at depths exceeding 500 fathoms, is best known in the northern part of the Atlantic and the seas corresponding with it (North Sea, Greenland Sea, Barents Sea). The species of the genus *Nymphon*, which occur in the neighbourhood of the coast of New England, are found to the north and east as far as Greenland, Spitzbergen, and Novaja Semlja ; but these Arctic Seas are, moreover, inhabited by some forms of the same genus occurring there only. As this point has been more fully discussed by me in another paper, it will suffice merely to mention it here.

Among the Pycnogonids of the Challenger Expedition, *Colossendeis megalonyx*, Hoek, is the only species, which, though found at a depth of from 55 to 120 fathoms, has a wide range ; about the 58th south parallel it was dredged off Kerguelen Island, and also between Patagonia and the Falkland Isles.

With respect to the true deep-sea species the material is by no means sufficient for the study of their geographical range. Of the thirty-six species of Pycnogonids brought home by the Challenger, nineteen are true deep-sea species. Of these only three belong to the northern hemisphere, viz., *Colossendeis minuta*, Hoek, south of Halifax ; *Phoxichilidium oscitans*, Hoek, west of the Azores ; and *Phoxichilidium mollissimum*, Hoek, off Yeddo; they were only dredged once and were new to science. Of the remaining sixteen, which belong to the southern hemisphere, one was dredged at lat. 65° 42′ S. (*Nymphon meridionale*, Hoek) and one almost under the equator (*Nymphon per-*

lucidum, Hoek, at lat. 0° 48′ S.). The fourteen remaining species were all dredged between lat. 33° 31′ S., and lat. 53° 55′ S.; and it is a remarkable fact, that those two latitudes limit a zone of about 20°, in which Pycnogonids seem to be rather common. However, even in this zone they are again much localised, being almost in every instance from near the coast of an island or continent. For miles the dredge was let down without bringing up a single specimen; whereas six species were found occurring at Stations 146 and 147, off the Crozets Islands (these Stations being very near to one another, 1 take as one); one at Station 157; three east of New Zealand, at Stations 168 and 169; five between Juan Fernandez and Valparaiso (Stations 298, 299, and 300); and two east of Buenos Ayres (Stations 320 and 325).

These facts indicate, I think, that the number of places inhabited by deep-sea Pycnogonids are not very numerous, and that where Pycnogonids do occur, many forms are, as a rule, found living together. This also ought to be observed when considering that between Stations 237 (off Yeddo) and 298 (between Juan Fernandez and Valparaiso), throughout a course of 11,775 miles, the dredge was let down sixty times and not a single Pycnogonid was obtained. This of course may partly be ascribed to the circumstance, that on an average the depth of that part of the ocean is too considerable to be inhabited by Pycnogonids; but as the depth at many stations during that part of the voyage did not exceed the depths of other stations at which Pycnogonids were dredged, this cannot be considered as the only reason. Also when the same circumstance is found to be the case in that large part of the South Atlantic between the Azores and Station 146, where during a course of more than 9000 miles the dredge was let down at 76 stations without a single return of Pycnogonids, and this although the depth at these stations is less, and at most of them much less, than some of the greater depths at which Pycnogonids were found, it is quite evident that the depth of the sea alone cannot be held responsible for it. Nor do I consider it yet proved that Pycnogonids are totally wanting in these oceans, as only a very small area of these oceanic abysses has been explored; so I think the only conclusion which at present may be drawn is this, that as yet only a few of the places where Pycnogonids occur in great numbers have been found out.

In regard to the nature of the bottom from which the Pycnogonids of the Challenger Expedition were obtained, conclusions must be also somewhat uncertain. The bottoms on which they occur seem to be extremely different. We find that one species was brought up from a bottom of gravel and stones, one from hard ground, one from rocky ground, five species are recorded as having been brought up from a muddy bottom, one from diatom ooze, five from a sandy bottom, three from a bottom of grey ooze, three from grey mud, and three from globigerina ooze. The other species were obtained from rocky bottoms in the neighbourhood of the shore.

More particulars about the geographical and bathymetrical distribution of the

Pycnogonids may be found in the list of the species hitherto described, which I append to this report. I have tried to make it as complete as possible; yet it contains many species of which no information is given as to the depth at which they were found; others of which even the locality they inhabit is not accurately stated; and furthermore, there are genera and species—and of the latter no small number—about which we are totally left in the dark. To explain this the reader must keep in mind (1) that this is the first

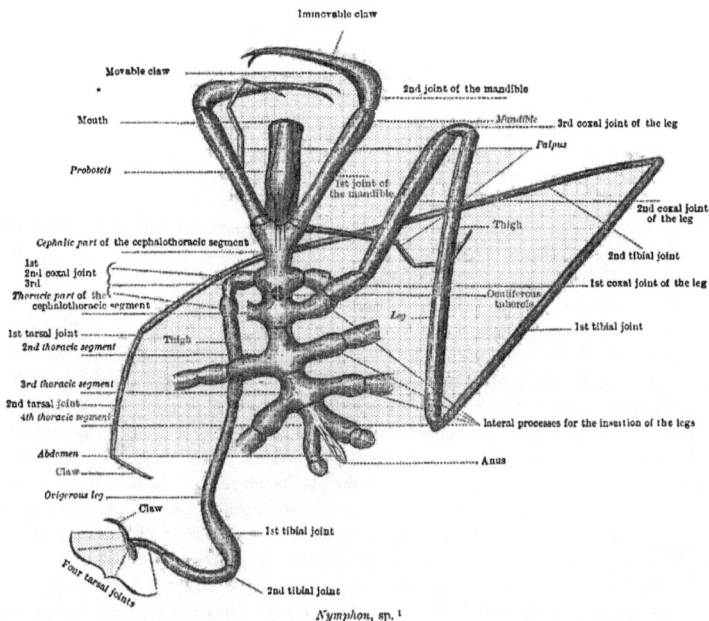

Nymphon, sp. [1]

attempt to make such a list, with the exception of a very incomplete and superficial enumeration published in 1874 by Semper,[2] and (2) that there is as yet no paper published which discusses the relative value of distinguishing marks. So it is evident that the making of this list has been an exceedingly troublesome affair, and that some allowance may be made for its incompleteness.

[1] For reasons easily to be understood I have taken a species of *Nymphon* as the type.

[2] C. Semper, Ueber Pycnogoniden und ihre in Hydroid-Polypen schmarotzenden Jugendformen. Arbeiten des Zool. Zoot Instituts in Würzburg, Band i., 1874.

Before inserting this list I wish to give a short description of the body of a Pycnogonid, and at the same time to state the nomenclature I have made use of.

The body of every Pycnogonid consists of four segments, the first of which is to be considered as formed by the connection of the head with the first thoracic segment. At the anterior end this first segment is furnished with a long and stout proboscis. This proboscis is situated either about the front of the first segment, as in *Nymphon*, and in this case is capable of very limited motion, or as in *Ammothea* and *Ascorhynchus*, though also situated about the front, it is connected with the segment by means of an articulation, and for that reason is highly movable, or it is, as in *Phoxichilidium*, situated on the ventral surface of the first segment, and bent forward; or finally, it is situated about the ventral side, and at the same time lapped over it (*Böhmia*, mihi). The form and size of this proboscis varies greatly. . At its extremity it is furnished with a triangular mouth. It is to be considered as an unpaired outgrowth of the region surrounding the mouth, and has nothing to do with a true head, as was supposed by Savigny. Neither is there anatomically or embryologically any real ground for the opinion, suggested by Huxley,[1] that the proboscis represents the united chelicerae and pedipalpi like that of Acarina.[2]

The cephalic part of the cephalothoracic segment is generally furnished with three pair of appendages, which long ago received the names of mandibles, palpi, and ovigerous legs. As far as has been ascertained till now, there is not a single genus of Pycnogonid, which does not show these three pair of appendages either in the adult state, or during its embryological development. Yet cases are not rare, in which in the adult animal, either the first (the mandibles) or the second pair (the palpi) or both are deficient. With respect to the third pair of appendages (the so-called ovigerous legs), on the contrary, they are never found wanting, as far as we know, in the adult animal of either sex. Whoever studies different forms of Pycnogonids, will soon discover what a difference may be caused in the appearance of the cephalic part of the body by the presence or absence of the cephalic appendages; hence it is that the various authors who have proposed a classification of the group have largely made use of this presence or absence of cephalic appendages. Although there is no doubt, I believe, that good characteristics may be derived from the number of these appendages, the following may show how extremely necessary it is to be cautious in this matter.

[1] Huxley, Anatomy of Invertebrated Animals, p. 386, London, 1877.
[2] On a transverse section, the proboscis of the Pycnogonids always shows a more or less distinctly triangular shape, the mouth is also triangular, &c. The total form, therefore, is to be compared with the fruit of a monocotyledonous plant, composed of three carpels. Of these one is placed dorsally, the two others meet longitudinally in the middle of the ventral side. If anybody should feel inclined to try again to homologise the proboscis with cephalic appendages, he will have to call the dorsal piece the labrum, and the two others the homologues of mandibles. However, in the earliest stages of development I have observed, the proboscis has already the form of a short cylindrical appendage, and I must point out the anatomical fact that the proboscis for the greater part is innervated from the supraoesophageal ganglion.

The mandibles in some genera are two-jointed (*Nymphon*, *Pallene*, &c.), in others three-jointed (*Phoxichilidium*). As a rule the second or third joint terminates in a pair of pincers, with a movable and an immovable claw. Now there are genera, some species of which show the mandibles small, yet furnished with true pincers, whereas other species of the same genus show the mandibles in a much more rudimentary state, as if, for instance, represented only by a single joint terminating abruptly (*Ascorhynchus glaber*, Hoek, and *A. minutus*, Hoek). In other genera the mandibles are in the adult animal always rudimentary, represented only by short stumps (*Lecithorhynchus*, Böhm, *Oorhynchus*, Hoek, &c.); whereas in a fourth category the mandibles have totally disappeared (*Colossendeis*, *Phoxichilus*, *Pycnogonum*, &c.). Among the specimens of a species of one of these genera (*Colossendeis gracilis*, Hoek), dredged during the cruise of H.M.S. Challenger, I have, however, found one specimen furnished with a pair of distinctly three-jointed mandibles, terminating in a pair of pincers; and this specimen was the largest of the three obtained.

The palpi when present show very different numbers of joints. Thus there are only three in *Pephredo*, five in *Nymphon* and *Discoarachne*, eight in *Achelia*, nine in *Ammothea* and *Corniger*, ten in *Ascorhynchus*, *Colossendeis*, &c. The palpi have disappeared in the genera *Pallene*, *Phoxichilidium*, *Phoxichilus*, *Pycnogonum*, &c. In *Phoxichilidium* they are as a rule still represented by rounded lateral processes placed at both sides of the front part of the cephalothorax, whereas Böhm has observed a specimen of *Pallene* furnished with rudimentary, yet distinct two-jointed palpi.

The third pair of appendages, viz., the ovigerous legs, are never wanting in any species of Pycnogonids. Among the Pycnogonids of the Challenger, there is not even a single specimen without ovigerous legs! As a rule they are ten-jointed; the first three joints are extremely small, the two following are the longest of all, the sixth joint is a great deal shorter, the last four joints are much shorter still, the tenth joint as a rule is furnished with a claw. In some genera (*Collossendeis*, e.g.) the fifth joint is small, the sixth as long as the fourth joint. In those genera, where a certain tendency is observed to drop their cephalic appendages, the ovigerous legs share this fate only to a small extent. As the functions of the ovigerous legs are twofold, one being to bear the eggs, a function only accomplished by the male,[1] the other to serve as an organ of feeling, also, in all probability, of seizing the food, and as the latter of these functions is almost identical with that of the other cephalic appendages, it is quite natural, I believe, that, whereas the males are never seen without these appendages, they are wanting in the females only of those genera which have also lost their other cephalic appendages. Finally, it is evident, that the males of those latter genera ought to show the ovigerous legs in such a rudimentary state, as to be fit only for the ovigerous function.

Dorsally the front part of the cephalothorax bears the oculiferous tubercle; although

---

[1] Hereafter I will show that this rule admits of an exception. See under *Nymphon brevicaudatum*, Miers.

in many instances—especially in the true deep-sea species—the eyes are wanting (a matter to be discussed hereafter), it never happens that the tubercle has totally disappeared. Most genera have this tubercle placed nearly in the middle between the two ovigerous legs; but in some genera (*Phoxichilidium, e.g.*) it is situated much nearer the front of the segment.

The thoracic part of the cephalothorax and the three following true thoracic segments are furnished with lateral processes for the insertion of the legs; these lateral processes in the different genera, and even in different species of the same genus, are of very different lengths. The segments of the body themselves are also of very different lengths. There are extremely slender forms with long segments and widely separated lateral processes, and there are also forms so highly concentrated, that the lateral processes are not separated at all; and between these extremes, which are often met with in one and the same genus, numerous intermediate forms are to be observed. The dorsal surface of the body is either smooth or furnished with knots, spines, strong prickles, &c.

At its extremity, between the two lateral processes for the insertion of the last pair of legs, the last thoracic segment has a rudimentary abdomen of varying length, which is sometimes (*Colossendeis, e.g.*) connected with the segment by means of an articulation, and also sometimes (*Zetes*, Kröyer) shows traces of being divided into two segments.[1] At its extremity the anal aperture is found.

The legs begin at the ends of the lateral processes; they are eight-jointed. For the joints I retain the names proposed by Johnston; these names are the same as those used in entomology, but it is evident that in this case identity of name does not necessarily go along with identity of meaning; neither analogical nor homological comparison is meant by it.

The first three (the coxal) joints are as a rule very short; the following three, the thigh and the two tibial joints, are much longer (the second tibial being in most cases the longest of all). The two tarsal joints are again a great deal shorter. The first tarsal as a rule is shorter than the second; in many instances it is even extremely small, its function then being only to furnish a highly movable articulation to the last joint of the leg. At its extremity the last joint is furnished with a claw, which is, or is not, accompanied by two accessory claws.[2] In some genera (*Colossendeis, e.g.*) accessory claws are never observed, while in other genera (*Nymphon*) they occur in some species and are wanting in others. Therefore the presence or absence of accessory claws alone should not be made use of in establishing new genera.

---

[1] *Rhynchothorax mediterraneus*, Cos., Microdoride mediterranea, Napoli, 1861, has a seven-jointed abdomen (Addome angusto e brevissimo di 7 articoli).

[2] I think there is not a single reason for calling this claw a ninth joint. At any rate the homology of the claw with its accessory claws is much greater than that between the claw and the joints of the leg, and, therefore, if the claw is considered as a ninth joint in those cases where accessory claws are observed, we must speak of a joint having two lateral joints close to its origin, which would be absurd.

CATALOGUE OF THE SPECIES OF PYCNOGONIDA AT PRESENT KNOWN, WITH INDICATIONS
OF THE HABITAT AND RANGE IN DEPTH OF EACH SPECIES.

A point of interrogation has been placed before uncertain or not sufficiently determined species, and an asterisk before those dredged by the Challenger Expedition, and during the cruise of the " Knight Errant"; of these a full description is given hereafter. In the left-hand column the range in depth of each species is given.

Class PYCNOGONIDA, Latr.

Crustacea haustellata, Johnston; *Crustacés aranéiformes* M.-Edw.; *Podosomata,* Leach; *Pantopoda,* Gerstæcker.

Family I. NYMPHONIDÆ

Contains those Pycnogonida which have both mandibles and palpi strongly developed. The ovigerous legs are always present in both sexes, and are, as a rule, furnished with denticulate spines. The only genus: *Nymphon.*

Depth in Fathoms.	Name.	Geographical Distribution.
	Nymphon. Fabr. (1794) Mandibles biarticulate, cheliform; palpi, five-jointed; ovigerous legs, ten-jointed.	
	A. Species with auxiliary claws.	
10 to 15 (Miers). 33 (Bell). 48 to 50 (Wilson). 52 (U. S. Fish. Com.). 299 (Sars).	*Nymphon hirtipes,* Bell, Belcher, The Last of the Arctic Voyages, vol. ii. p. 408, 1855. Prof. G. O. Sars (1877) considers this species to be the same as *N. hirtum,* Fabr., which in that case would not be identical with *N. hirtum,* Fabr., as described by Kröyer (1845). As it is impossible to recognise the species by the description of Fabricius, I think it safest to retain the name *N. hirtum* for the species of Kröyer whose description was published long before Sars' Prodromus. *N. hirtum,* Fabr., Sars, in Prodromus descriptionis, &c. (Arch. für Math. og Nat., ii. 1877). *N. hirtipes,* Bell, Wilson, Pycnogonida of New England, Trans. Connect. Acad., vol. v., 1880. Having received from the U. S. Fish. Commission a specimen brought up off Halifax, I feel certain that the animal described by Wilson belongs to this species. Finally, I believe that the specimens studied by Miers, and referred by him to *N. hirtum* (Arctic Crustacea in Ann. and Mag. Nat. Hist., 4th series, xx. 108, 1877) belong to this species and not to *N. hirtum,* Fabr. The *N. hirtum,* var. *obtusidigitum,* Miers, seems to be an undeveloped female of the same species.	Lat. 64° 36′, long. 10° 21′ 5″. Off Halifax; Franklin-Pierce Bay; Discovery Bay; Floeberg Beach; Northumberland Sound. " Appears to be a common inhabitant of the high northern latitudes " (Miers). (A common species at different stations in the Barents Sea. Hoek, in MS.)

K 3

Depth in Fathoms.	Name.	Geographical Distribution.
80 to 90	*Nymphon pallenoides*, G. O. Sars, Crustacea et Pycnogonida nova, Arch. f. Math. og Naturvid., iv., 1879, p. 470. Of this species Sars says that it is *N. hirsuto* affine, sed statura minore, &c. Perhaps it is only a local variety. As Sars does not mention the length of the auxiliary claws, I feel uncertain whether it is nearest to *N. hirtipes*, Bell, or *N. hirtum*, Fabr. (Kröyer).	Coast of Norway (Saltenfjord).
	Nymphon hirtum, Fabr. Entom. Systematica, iv. 417, 1794. *N. hirtum*, Chr. Fabr. (?), Kröyer, Bidrag, &c., Nat. Tidskr., N. R. i. 113, 1845. The description of Fabricius being quite insufficient, I retain the species with the diagnosis of Kröyer. Perhaps *N. hirsutus*, Sabine (Appendix, &c., p. ccxxvi, 1824), belongs also to this species; according to Kröyer this is doubtful. *N. hirtum*, O. F. Buchholz, Crustaceen der Zweiten Deutschen Nordpolarfahrt, 1874, ii. 397. In regard to *N. hirtum*, Fabr. Heller (Crust., Pycnogon., und Tunicaten der K. K. Ost.-Ung. Nordpol. Exp. in Denkschr. d. Kais. Akad. d. Wiss., Bd. xxxv., 1875), it is impossible to determine whether it is this species or *N. hirtipes*, Bell, that was observed.	Norwegian Ocean, Iceland, East Coast of Greenland (Böhm, Pycnogoniden des Museums zu Berlin, 1879) ; East Coast of Greenland (Nordshannan), Storfjord (Spitzbergen), Buchholz, *loc. cit.*
15 to 65 (Böhm). 25 to 120 (Challenger).	*Nymphon brevicaudatum*, Miers, Ann. and Mag. of Nat. Hist., 4th series, vol. xvi. p. 117, 1875). Crustacea of Kerguelen Island, Phil. Trans. Lond. vol. clxviii. Extra vol., pp. 200–214, 1879, pl. xi. fig. 8. *N. horridum*, Böhm, Pycnogoniden des Museums zu Berlin, Monatsber, der Königl. Acad. d. Wiss. zu Berlin, 1879, p. 175, taf. i. fig. 3–3f. Böhm's supposition that his *N. horridum* was indentical with Miers' *N. brevicaudatum* is true, as has been proved by the more extensive description with illustrations published by Miers in 1879 in the extra vol. of the Phil. Trans. of London. Many specimens of this species were obtained during the visit of H.M.S. Challenger to Kerguelen.	Kerguelen.
10 to 80 (Miers). 35 to 90 (Wilson). 515 to 540 (" Knight-Errant "). 110 to 160 (Barents Sea, Hoek in MS.).	*Nymphon strömii*, Kr., Nat. Tidskr., N. R., vol. i. p. 111, 1845. Wilson (Pycnogonida of New England, Trans. Connect. Acad., vol. v. p. 17, 1880) believes that the *N. gracilipes*, Heller (Crust. Pycnog. und Tunic. des K. K. Oester.-Ungar. Exped. Denkschr. d. K. Ak. der Wiss., xxxvi., 1875)—not to be confounded with the *N. gracilipes*, Miers, and therefore named by Böhm (Pycnogoniden des Museums zu Berlin, Monatsb. der K. A. d. Wiss. zu Berlin, p. 170, 1879) *N. Helleri*—is very closely allied if it is not identical with this species. This is also my opinion, although Heller in his	Coast of Norway ? (Kröyer), North Atlantic (Sars and " Knight Errant " cruise), Barents Sea (Hoek in MS.). " An verschiedenen Punkten " (Heller, *i.e.*, during the Austria-Hungarian North Polar Exped. of 1873) ; Floeberg Beach, Cape Fraser, Grinnell-land (Miers), Coast of North America, Gulf of St Lawrence,

Depth in Fathoms.	Name.	Geographical Distribution.
(?)	diagnosis of his species says : unguiculi auxiliares nulli (p. 40). A considerable number of specimens of this species have recently been dredged (North of Scotland) during the cruise of the " Knight Errant." One of the commonest species in the northern part of the Atlantic. (?) *Nymphon giganteum*, Goodsir, Ann. and Mag. of Nat. Hist., vol. xv. p. 293, 1845. Goodsir's description is not sufficient to determine this species. Wilson (Pycnogonida of New England, Trans. Connect. Acad., vol. v. p. 16, 1880) considers this species as identical with the *N. strömii* of Kröyer. As Goodsir does not even say whether the body is slender or robust, it is extremely difficult to ascertain whether or not Wilson's suggestion is right.	off the Isles of Shoals, off Halifax (Wilson). Sea at Embleton, Northumberland.
412 540 (" Knight Errant")	*Nymphon macronyx*, G. O. Sars, Prodromus, Arch. f. Math. og Naturvid., ii. 365, 1877. I know this species by the description of Prof. Sars, from a pencil-drawing he kindly sent me, and feel sure it is a good species. Recently I had a fair opportunity of becoming acquainted with it, as numerous specimens were dredged north of Scotland, and forwarded to me by Mr Murray. As only a very short diagnosis of this species has been given by Prof. Sars, I will publish hereafter a detailed description of it.	Lat. 62° 44′ 5″ N., long. 1° 48′ E. To the north of Scotland.
83	*Nymphon brevicollum*, n. sp. A fine, well-characterised species.	South of Halifax, Station 49, Challenger Expedition.
45 to 120	*Nymphon brachyrhynchus*, n. sp. A small but well-characterised species, which seems to abound at a depth of 45 to 120 fathoms.	Kerguelen.
1675	*Nymphon meridionale*, n. sp., shows a certain affinity to *N. gracilipes*, Miers (Ann. and Mag. of Nat. Hist., 4th ser., vol. xvi. p. 76, 1875) ; so the latter species may be considered as a shore-relation of this true deep-sea species.	South of Kerguelen Island, lat. 65° 42′ S., long. 79° 49′ E., Station 43, Challenger Expedition.
10 to 45	*Nymphon gracilipes*, Miers, Ann. and Mag. of Nat. Hist., 4th series, vol. xvi. p. 76, 1875. *N. gracilipes*, Miers, Böhm, Pycn. des Mus. zu Berlin, Monatsb. der K. Akad. der Wiss. zu Berlin, p. 170, 1879. *N. antarcticum*, Miers, Crustacea of Kerguelen Island, Phil. Trans. Lond., vol. clxviii. Extra vol., p. 200–214, pl. xi., 1879. This species seems to be a good species. Miers thought it necessary to alter the name he originally proposed, because in the same year the same name was given by Heller to an arctic species. The *N. gracilipes*, Heller, being only a synonym of the *N. strömii*, Kr., I think it best to retain this name for the Kerguelen species, as was originally proposed by Mr Miers.	Kerguelen.

Depth in Fathoms.	Name.	Geographical Distribution.
120	*Nymphon fuscum, n. sp. This species too seems to be closely allied to N. gracilipes, Miers.	Kerguelen.
2225	*Nymphon longicollum, n. sp. The only specimen of this true deep-sea species shows very characteristic features.	Off Coast of Chili (Station 298, Challenger Expedition).
1 to 2	Nymphon gracile, Leach, Zool. Misc., vol., i. p. 45, 1814. N. gracile, Leach, Johnston, An Attempt, &c., in Mag. of Zool., 1837. N. gracile, Leach, Hoek, Ueber Pycnogoniden, Nied. Arch. f. Zool., iii., 1877. It has been suggested by Kröyer that it might be the same as N. grossipes, O. Fabr., but I prefer to retain the name of Leach for the species as known by the description and figures of Mr Johnston and myself, which is distinct from N. grossipes, O. Fabr., as described by Kröyer.	British Seas everywhere (Leach), Dutch Coast (Texel) (Hoek).
229 ; 417	Nymphon megalops, G. O. Sars (Prodromus descriptionis, &c., Arch. f. Math. og Naturvid., ii. 366, 1877). Not figured. Prof. Sars kindly sent me a pencil-drawing, and from this drawing and his Latin diagnosis I believe the species is nearly related to N. gracilipes, Miers.	Lat. 63° 10′ 2″ N., long. 4° 59′ 6″ E. Lat. 64° 36′ N., long. 10° 21′ 5″ E.
825	*Nymphon perlucidum, n. sp. A very well-characterised species, of which, unfortunately, only one specimen was dredged by the Challenger.	Between Celebes and Halmahera.
60 (Wilson) 220 (Sars) 67 (Hoek in MS.).	Nymphon longitarse, Kr., Bidrag till Kundskab, Natur. Tidskr. N. R., i. 112, 1845, is so nearly related to N. mixtum, Kr., that it may, perhaps, be only a variety of that species, and in that case, of course, of N. grossipes, O. Fabr. ; however Wilson (Pycnogonida of New England, Trans. Connect. Acad., vol. v. p. 19, 1880) believes it a good species, readily distinguished by its extremely attenuated appearance. From the Barents Sea I got some specimens, which unquestionably belong to the form described by Kröyer.	Coast of Greenland and West Norway ; off Halifax, St George's Banks, lat. 61° 47′ 2″ N., long. 3° 18′ 5″ E. Barents Sea (Hoek in MS.).
Shore ; " majores etiam in profundo" (Oth. Fabr. Fauna Groenl) 50,20 to 100 (Wilson) 83 (Challenger). 540 ("Knight Errant".	*Nymphon grossipes, Oth. Fabr. (sp.), 1780. Pycnogonum grossipes, Oth. Fabr., Fauna Groenlandica, p. 229, 1780. (?) Phalangium grossipes, Linn., Syst. Naturæ, xii. 1027, 1766 (?). (?) Nymphon grossipes, Fabr., Entom. System. emendata et aucta, tom. iv. p. 417, 1794 (?). N. grossipes, Oth. Fabr., Kröyer, Bidrag till Kundskab, Nat. Tidskr., N. R., i. 108, 1845. N. grossipes, Oth. Fabr., Wilson, Pycnogon. of New England, Trans. Connect. Acad., v. 21, 1880. The species is best known from the descriptions of Kröyer and Wilson. Three specimens were obtained during the Challenger Expedition, and a single	North Sea (Böhm, Pycnogoniden des Museums zu Berlin, Monatsber. der K. Akad. d. Wiss. zu Berlin, 1879) ; Coast of Norway (Kröyer), Barents Sea (Hoek in MS.), Northern part of the North Atlantic (" Knight Errant," 1880) ; East Coast of Greenland (Fabricius, Buchholz) ; North Georgian Islands (Sabine, Suppl. to the Ap-

Depth in Fathoms.	Name.	Geographical Distribution.
	one during the cruise of the " Knight Errant " north of Scotland.	pendix, &c., p. ccxxv. 1824); Coast of North America, from the Gulf of St Lawrence, as far south as Long Island Sound (Wilson), Challenger Exped., Station 49, South of Halifax, U. S. A.
417 (Sars).	(?) *Nymphon mixtum*, Kr., Nat. Tidskr. N. R., i. p. 100. 1845. *N. mixtum*, Kr., Buchholz, Zweite Deutsche Nordpolarfahrt, Crust., p. 397, 1874. *N. mixtum*, Kr., G. O. Sars, Prodromus (Archiv. f. Math. og Naturvid., ii. 366, 1877) = *N. grossipes*, O. Fabr., Wilson, Pycnogonida of New England, Trans. Connect. Acad., vol. v. p. 20, 1880. Wilson thinks the *N. mixtum*, Kr., is undoubtedly a form of *N. grossipes*, O. Fabr. I also believe it so nearly related to *N. grossipes*, that it may be only a variety. Buchholz writes (*loc. cit.*, p. 397):—" Doch muss ich es dahingestellt sein lassen, ob die von Kröyer angegebene auf dem Verhältniss der Länge des Tarsus zum Endglied beruhende Artunterscheidung ausreichend ist, um diese Art von der vorigen (*N. grossipes*, O. Fabr.) zu trennen."	Coast of West Norway (Kröyer), lat. 63° 10' 2" W., long. 4° 59' 6" E. (Sars); East Coast of Greenland, Spitzbergen (Buchholz, *loc cit.*).
(?)	(?) *Nymphon brevitarse*, Kr., Nat. Tidskr. N. R., i. 115, 1845. *N. hirsutum*, Kr., Grönlands Amfipoder, p. 92, 1838 (Autoritate, Kröyer) = *N. grossipes*, O. Fabr., Wilson, Pycnogonida of New England, Trans. Connect. Acad., p. v. 20, 1880. In all probability Wilson is right when he says *N. brevitarse* is only a form of *N. grossipes*.	Coast of Greenland.
160 (Hoek in MS).	*Nymphon sluiterii*, Hoek, in MS., Pycnogonids of the first two cruises of the W. Barents. A well-characterised arctic species, with extremely small auxiliary claws, second joint of the palpus longer than third, first tarsal joint longer than second, with a claw at the end of the legs, which is not shorter than the last joint of the leg, and a truncate oculiferous tubercle.	Barents Sea (Hoek in MS.).
146 to 180 (G. O. Sars). 160 (Hoek in MS.).	*Nymphon serratum* G. O. Sars, Crustacea et Pycnogonida nova, Arch. f. Math. og Naturvid., iv., 471, 1879. An extremely characteristic species, with a large spine dorsally on the first three segments of the trunk.	South of Spitzbergen, Barents Sea (Hoek in MS.).
(?) 25 to 35	(?)*Nymphon brevirostris*, Hodge, Mennell, Report on Dredging off the Northumberland Coast and Doggerbank, British Association Report, p. 119, 1862. A very unsatisfactorily described species; in all probability the same as *N. brevitarse*, Kr. = *N. grossipes* O. Fabr.	(?) Northumberland Coast, Doggerbank.

Depth in Fathoms.	Name.	Geographical Distribution.
(?)	(?) *Nymphon glaciale*, Lilljeb., *N. glaciale*, Lilljeb., Jarzynsky, Præmissus Catalogus Pycnogonidarum in mari glaciali, Annales de la Société des Naturalistes de St Petersbourg, 1870. I have never seen a specimen of it, nor do I know where the description of Lilljeborg is to be found.	White Sea.
(?)	(?) *Nymphon femoratum*, Leach, Zool. Misc., i. 45, pl. xix. fig. 2, 1814. *N. femoratum*, Leach, Milne-Edwards, Hist. Nat. des Crustacés, iii. 534, 1840. *N. femoratum*, Leach, Johnston, An Attempt &c., Mag. of Zool. and Botany, i., 1837. In all probability not a good species : description very insufficient ; dilated thighs are common to the females of almost all the species.	Channel.
(?)	(?) *Nymphon pellucidum*, Goodsir, Edin. New Phil. Journ., vol. xxxii., 1842. Characterised very insufficiently : may turn out to be a variety of *N. brevitarse*, Kr.	Coast of England.
(?)	(?) *Nymphon spinosum*, Goodsir, Edin. New Phil. Journal, vol. xxxii., 1842. Like the other species of Goodsir, *N. spinosum* has been described so very insufficiently that it is not to be recognised.	Coast of England.
(?)	(?) *Nymphon johnstonii*, Goodsir, Edin. New Phil. Journal, vol. xxxii., 1842. A very uncertain species.	Coast of England.
(?)	(?) *Nymphon minutum*, Goodsir, Edin. New Phil. Journal, vol. xxxii., 1842. Goodsir's description is so insufficient that the species is not to be recognised.	Coast of England.
(?)	(?) *Nymphon longiceps*, Grube, 46ster Jahres-Ber. der Schles. Gesellsch. f. vaterl. Cult. p. 54, 1869. The description of this species is so insufficient that it is impossible to recognise it.	China Sea.

B. Species without auxiliary claws.

480 (Station 56, "Porcupine"), 412 : 299 (Sars). 375, 540 ("Knight Errant"). 120 to 160 (Barents Sea, Hoek in MS.).	* *Nymphon robustum*, Bell, Belcher's Last of the Arctic Voyages, vol. ii. p. 409, 1855, Tab. xxxv. fig. 4 = *N. hians*, Heller, Crust, Pycnog. und Tunicaten der K. K. Oester. Ungar. Nordpol. Exped., Denkschr, d. Wiener Akad. xxxv. p. 41, 1875, = *N. abyssorum*, Norm., Wyville Thomson, Depths of the Sea, p. 129, 1873. I quite agree with G. O. Sars (Prodromus, Arch. for Math. og Naturvid., ii. 365, 1877), who places *N. hians*, Heller, and *N. abyssorum*, Norman, as identical with this species. A large and excellent species abounding in the higher northern latitudes. The largest haul of Pycnogonids, Mr Murray writes to me, he ever saw was that in which he got an immense number of specimens of this species (Cruise of the "Knight Errant" to the north of Scotland, Aug. 1880).	60° 2′, 62° 44′ 5″, 64° 36′ N. lat. ; 6° 11,′ 1° 26′ W. long., 1° 48′ E. long. Barents Sea (Hoek in MS.). North of Scotland.

Depth in Fathoms.	Name.	Geographical Distribution.
1100	*Nymphon compactum, n. sp. A well-characterised deep-sea species, of which only females—two specimens in all—have been dredged.	East of Auckland (Station 128, Challenger Expedition).
50	Nymphon phasmatodes, Böhm, Pycnogoniden des Museums zu Berlin, Monatsb. der K. Akad. der Wiss. zu Berlin, p. 173,1879,Taf. i. fig. 2–2b. Seems to be a good species.	Cape of Good Hope.
1375 to 1600	*Nymphon hamatum, n. sp. A good and common deep-sea species.	Between Kerguelen and Cape of Good Hope.
1100	*Nymphon longicoxa, n. sp. A fine well-characterised deep-sea species, of which thirteen specimens were dredged by the Challenger.	East of Auckland (Station 168, Challenger Expedition).
2160	*Nymphon procerum, n. sp. A slender deep-sea species, which is well-characterised, yet requires further study.	West of Valparaiso (Station 299, Challenger Expedition).

Family II. COLOSSENDEIDÆ

Contains those Pycnogonida which have either rudimentary mandibles or no mandibles at all, strongly-developed palpi and ovigerous legs present in both sexes, and as a rule furnished with denticulate spines. The genera belonging to this group are very numerous : *Ammothea*, *Ascorhynchus*, *Achelia*, and *Colossendeis* are typical representatives.

Depth in Fathoms.	Name.	Geographical Distribution.
	Ammothea, Leach (1815).	
	Mandibles biarticulate, cheliform, feeble ; palpi nine-jointed ; ovigerous legs ten-jointed ; proboscis pyriform.	
(?)	Ammothea carolinensis, Leach, Zool. Misc., vol. i. p. 34, 1815, pl. xiii. Of this species a very good drawing has been published by Leach.	South Carolina.
0 to 5	Ammothea pycnogonoides, Quatrefages, Mémoire sur l'organisation des Pycnogonides, Ann. d. Sc. Nat. 3ième série, Zool., tom. iv., Paris, 1845. I think this species is closely related to A. longipes, Hodge, but the descriptions of both authors are so very insufficient that this is not to be made out.	Saint Malo (Coast of France).
(?)	Ammothea longipes, Hodge, Ann. and Mag., 3d series, vol. xiii. p. 114, 1864 ; Grube, Verhandl. d. Schles. Gesellsch. f. vat. Cultur., 1869–72. The description of this species is also quite insufficient ; it cannot be made out whether the species described by Grube is identical with that of Hodge. The specimen of Grube is furnished with six-jointed palpi, whence I believe it not to have been an adult animal. In all probability it is the same as that described by Quatrefages as A. pycnogonoides, Ann. d. Sc. Nat., iii. série, Zool., tom., iv., 1845.	Polperro (English Coast) (St Vaast la Hougue (Grube); Roscoff (Grube).

Depth in Fathoms.	Name.	Geographical Distribution.
(?) (Deep-water).	*Ammothea brevipes*, Hodge, List of the British Pycnogonoidea, with descriptions of several new species, Ann. and Mag. of Nat. Hist., 3d series, xiii. 113, 1864. Very insufficiently described.	Durham Coast; Heligoland (Semper).
(?)	*Ammothea achelioides*, Wilson, Transact. Connect. Acad., vol. v. part I, p. 16, 1880. "In general appearance it is closely similar to *Achelia spinosa*, Stimpson" (Wilson, *loc. cit.*). No doubt there is a near relation between the genera *Achelia* and *Ammothea*, yet a minute examination of "adult" specimens of both genera makes it necessary to consider these genera provisionally as distinct.	Bay of Fundy (U S. A.).

Böhmia, Hoek (1880).

Mandibles two-jointed, cheliform ; palpi seven-jointed ; ovigerous legs ten-jointed ; proboscis conical, entirely bent over to the ventral side.

(?)	*Böhmia chelata*, Böhm, sp.; *Pycnogonum chelatum*, Böhm, Pycnogoniden des Museums zu Berlin, Monatsb. der K. Akad. der Wiss. zu Berlin, 1879, p. 192 (pl. ii. fig. 5–5d). As the specimen observed by Böhm is furnished with ten-jointed ovigerous legs, I think there is not a single reason to consider it as a *Pycnogonum* in the larval condition. I therefore propose to form a new genus for it, and to call it after Mr Böhm.	(?)

Phanodemus, Costa (1836).

Mandibles cheliform ; palpi three- or four-jointed ; ovigerous legs (?) ; proboscis conical.

(?)	*Phanodemus horridus*, Costa, Fauna del regno di Napoli, Crostacei et Aracnidi, Napoli, 1838, p. 8. Description of this species extremely insufficient.	Gulf of Taranto.
(?)	*Phanodemus collaris*, Costa, Fauna del regno di Napoli, Crostacei et Aracnidi, Napoli, 1838, p. 8. Also described insufficiently.	Gulf of Taranto, Gulf of Naples.
(?)	*Phanodemus inermis*, Costa, Fauna del regno di Napoli, Crostacei et Aracnidi, Napoli, 1838, p. 9. Description not better than those of the foregoing species.	Gulf of Taranto, Gulf of Naples.

Rhynchothorax, Costa (1861).

Mandibles cheliform, four-jointed ; palpi eight-jointed ; ovigerous legs (?) ; proboscis long-ovate.

(?)	(?) *Rhynchothorax mediterraneus*, Cos., Microdoride mediterranea, Napoli, 1861, p. 8, Taf. i. fig. 102. A very curious animal, with a short and narrow abdomen of seven joints. The description, however, is insufficient.	North Coast of Africa.

Depth in Fathoms.	Name.	Geographical Distribution.
	Pephredo, Goodsir (1842).	
	Mandibles cheliform; palpi three-jointed; ovigerous legs six-jointed; proboscis short, cylindrical.	
(?)	*Pephredo hirsuta*, Goodsir, Edin. New Phil. Journal, 1842, vol. xxxii. p. 136. The description of this species as given by Goodsir is very insufficient.	Coast of England.
	Platychelus, Costa (1861).	
	Mandibles cheliform; palpi five-jointed; ovigerous legs (?); proboscis pyriform.	
(?)	*Platychelus sardonicus*, Costa, Microdoride mediterranea, Napoli, 1861, p. 11. The description of the genus and species is given from a single specimen, and this has been most probably an immature female.	Coast of Sardinia.
	Oiceobathes, Hesse (1867).	
	Mandibles small, three-jointed, cheliform; palpi eight-jointed; ovigerous legs (?); proboscis conical.	
(?)	(?)*Oiceobathes arachne*, Hesse, Ann. d. Sc. Natur., 5ᵉᵐᵉ série, vii. 201, 1867. Description quite insufficient. Perhaps the genus *Oiceobathes* is the same as *Ammothea*.	Coast of France (Brest).
	Ascorhynchus, G. O. Sars (1877).	
	Mandibles rudimentary, cheliform or not cheliform; proboscis pyriform, more or less bent over to the ventral side; palpi ten-jointed; ovigerous legs ten-jointed.	
1081 to 1539.	*Ascorhynchus abyssi*, G. O. Sars, Arch. f. Math. og Naturvid., ii. 367, 1877. This well-characterised species is (according to Sars) very common in the great depths of the cold region.	Atlantic between Iceland and Norway. Lat. 63° 7′ N., long. 1° 26′ W.; lat. 68° 13′ 5″ N., long. 0° 33′ E.
1375	*Ascorhynchus glaber*, n. sp. A beautiful and large *Ascorhynchus*: not very different from *A. abyssi*, G. O. Sars, Arch. f. Math. og Naturvid., ii. 367, 1877.	Between Cape of Good Hope and Kerguelen.
(?)	*Ascorhynchus ramipes*, Böhm (sp.). *Gnamptorhynchus ramipes*, Böhm, Sitzber. der Ges. Naturf. Freunde zu Berlin, 1879, p. 56. *Ibid.*, p. 140.	Enosima (Japan).
38	*Ascorhynchus minutus*, n. sp. A well-characterised species, nearly allied to *A. ramipes*, Böhm.	Southern Coast of Australia: off entrance to Port Philip.
130	*Ascorhynchus orthorhynchus*, n. sp. A very fine and, I believe, well-characterised species.	Great Ocean North of Admiralty Islands.
	Zetes, Kröyer (1845).	
	Mandibles rudimentary, three-jointed; palpi ten-jointed; ovigerous legs nine- (nonne ten- (?)) jointed; proboscis long-ovate.	
(?)	*Zetes hispidus*, Kröyer, Naturhist. Tidskr. Ny Raekke, i. 117, 1845. *Zetes hispidus*, Kröyer, Jarzynsky, Praemissus Catalogus, Annales de la Soc. des Natur. de St Petersbourg, 1870.	Coast of Greenland, Coast of Russian Lapland.

Depth in Fathoms.	Name.	Geographical Distribution.
	Parazetes, Slater (1879).	
	Mandibles rudimentary, two-jointed ; palpi nine-jointed ; ovigerous legs ten-jointed ; proboscis fusiform.	
(?)	*Parazetes auchenicus*, Slater, Annals and Mag. of Nat. Hist., 5th series, iii. 281, 1879. Slater calls the generic characteristic of *Parazetes* "distinct." The description of the genus and species has been given from a single specimen, so I doubt whether the generic differences from *Zetes* or any other closely-allied genus will hold good.	Off Cape Sima (Japan).
	Pariboea, Philippi (1843).	
	Mandibles rudimentary, two-jointed ; palpi five-jointed ; ovigerous legs nine (nonne ten- (?)) jointed ; proboscis ovate.	
(?)	*Pariboea spinipalpis*, Philippi, Arch. f. Naturg., 1843, ix. p. 178. Perhaps the description of this species is not taken from an adult specimen.	Sorrento (Gulf of Naples).
	Alcinous, Costa (1861).	
	Mandibles rudimentary, three-jointed ; palpi seven- or nine-jointed ; ovigerous legs eight-jointed ; proboscis long-ovate.	
(?)	*Alcinous vulgaris*, Costa, Microdoride mediterranea, Napoli, 1861, p. 13. This species has, I believe, a certain resemblance to an *Ascorhynchus*.	Mediterranean.
(?)	*Alcinous megacephalus*, Costa, Microd. medit. Napoli, 1861, p. 14. This species belongs rather to *Pariboea*, Philippi, than to *Alcinous*, Costa.	Gulf of Naples.
	Achelia, Hodge (1864).	
	Mandibles rudimentary, two-jointed ; palpi eight-jointed ; ovigerous legs ten-jointed ; proboscis pyriform.	
Low water mark to 17 fathoms ; 35 fathoms.	*Achelia spinosa*, Stimpson (sp.). *Zetes spinosa*, Stimpson. Invertebrata of Grand Manan, 1853, p. 37. *A. spinosa*, Wilson, Pycnogon. of New England, Trans. Connect. Acad. of Arts and Sciences, vol. v. p. 7, 1880. The differences of *A. spinosa*, Stimpson, from Hodge's *A. echinata* "seem scarcely sufficient to separate them as distinct species" (Wilson, *loc. cit.*).	Grand Manan ; Eastport, Casco Bay, off Block Island, &c.
Low tide to a few faths. (Hodge).	*Achelia echinata*, Hodge, British Pycnogonoidea, Ann. and Mag. of Nat. Hist., 3d ser., vol. xiii. p. 115, 1864. A good description of this species has been given by Grube, St Vaast la Hougue und seine Meeres—besonders seine Anneliden-Fauna, Verhandl. d. Schles. Gesellsch. f. vaterl. Cultur, 1869–72, p. 27 (Separatabdruck). Perhaps identical with *A. spinosa*, Stimpson, Invertebrata of Grand Manan, 1853, p. 37.	Channel Islands, Isle of Man ; Durham Coast, St Vaast la Hougue, Dutch Coast.

Depth in Fathoms.	Name.	Geographical Distribution.
(?)	*Achelia lævis*, Hodge, British Pycnogonoidea, Ann. and Mag. of Nat. Hist., 3d ser., vol. xiii. p. 115, 1864. Böhm (Monatsb. d. K. Akad. der Wiss. in Berl., 1879, p. 186) refers to the same species a specimen from Kerguelen with seven-jointed palpi.	Polperro, Cornwall (Hodge), Nizza (Grube), Kerguelen (Böhm).
(?)	*Achelia hispida*, Hodge, British Pycnogonoidea, Ann. and Mag. of Nat. Hist., 3d ser., vol. xiii. p. 115, 1864. I doubt, whether this species described by Hodge is indeed a true species.	Polperro, Cornwall.

Tanystylum, Miers (1879).

Mandibles rudimentary, one-jointed; palpi six-jointed; ovigerous legs ten-jointed; proboscis rounded-conical.

5 to 7	*Tanystylum styligerum*, Miers, Crustacea of Kerguelen Island, Phil. Trans. London, vol. clxviii., extra vol., pp. 200–214, pl. xi., 1879. *Nymphon styligerum*, Miers (sp.), Ann. and Mag. of Nat. Hist., 4th ser., vol. xvi., 1875. Most probably this is a good species; however, I do not feel quite sure that the two specimens studied by Miers are really full-grown animals.	Kerguelen Island, Observatory Bay.
Down to 14 fathoms (Wilson).	*Tanystylum orbiculare*, Wilson, Pycnogonida of New England, Trans. Connect. Acad., vol. v. p. 5, 1880. Whether or not *Tanystylum* will prove a well-characterised genus, or will turn out to be identical with *Achelia*, Hodge, is doubtful as yet.	Vineyard Sound, and occurs as far south as Virginia (Wilson, *loc. cit.*).

Lecythorhynchus, Böhm (1879).

Mandibles rudimentary, two-jointed or represented by small knobs; palpi nine-jointed; ovigerous legs ten-jointed; proboscis cylindrical.

3 to 4 fathoms.	*Lecythorhynchus hilgendorfi*, Böhm, Monatsb. der K. Akad. d. Wiss. Berlin, 1879, p. 187. *Corniger hilgendorfi*, Böhm (sp.), *ibid.* and Sitzungsber. d. Ges. Naturf. Freunde zu Berlin, 1879, p. 140. Böhm has given a good description of this species, *loc. cit.*	Enosima (Japan).
(?)	*Lecythorhynchus armatus*, Böhm, Sitzungsber. der Ges. Naturf. Freunde zn Berlin, 1879, p. 140. *L. hilgendorfi*, Böhm, is furnished with rudimentary mandibles represented by small knobs; *L. armatus*, Böhm, with rudimentary but two-jointed mandibles. There can be no doubt, however (according to Böhm), that these species belong to the same genus.	Yeddo (Japan).

Oorhynchus, Hoek (1880).

Mandibles rudimentary, one-jointed; palpi nine-jointed; ovigerous legs ten-jointed; proboscis broadly ovate.

700	*Oorhynchus aucklandiæ*, n. sp. Of this very characteristic Pycnogonid, only a single specimen was dredged during the voyage of H.M.S. Challenger.	East of Auckland.

Depth in Fathoms.	Name.	Geographical Distribution.
	Colossendeis, Jarzynsky (1870). No mandibles; palpi ten-jointed; ovigerous legs ten-jointed; proboscis long, cylindrical, club-shaped or bottle-shaped.	
120 to 250 (Jarzynsky). 110 to 166 (Hoek in MS. 540 ("Knight Errant").	*Colossendeis proboscidea*, Sabine (sp.). *Phoxichilus proboscideus*, Sabine, Supplement to the Appendix of Captain Parry's Voyage, Zoology, p. ccxxvi. *C. borealis*, Jarzynsky, Præmissus catalogus Pycnogonidarum inventarum in mari glaciali ad oras Lapponicæ rossicæ et in mari albo, anno 1869 et 70, Annales de la Soc. des Natur. de St Petersb., 1870. *C. proboscidea*, Sabine, G. O. Sars, Prodromus descriptionis, &c., Arch. f. Math. og Naturv., ii. 268, 1877. A gigantic Pycnogonid of the higher northern latitudes.	Coast of Russian-Lapland, North Atlantic. Lat. 62° 44′ 5″ N., long. 1° 48′ E. Barents Sea. (Hoek in MS.), North of Scotland ("Knight Errant.").
55 to 120	*Colossendeis megalonyx*, n. sp. A well-characterised species. In some respects it resembles the *C. proboscidea*, Sabine, of the higher northern latitudes; it has, however, a much more slender body and longer legs.	Lat. 50° S.: Kerguelen; between Patagonia and the Falkland Islands.
417	*Colossendeis angusta*, G. O. Sars, Prodromus descriptionis, &c., Arch. f. Math. og Naturvid., ii. 268, 1877. This species is the slender *Colossendeis* of higher northern latitudes.	North Atlantic; lat. 63° 10′ 2″ N., long. 4° 59′ 6″ E.
400; 1375; 1600	*Colossendeis leptorhynchus*, n. sp. This species seems to be a very good one. It is easily recognized by its extremely long and narrow proboscis, of an almost cylindrical shape.	Between Kerguelen and Cape of Good Hope; West of Valparaiso; between Hannover Isle and Patagonia.
1800	*Colossendeis gigas-leptorhynchus*. A single specimen of this form was dredged at Station 158. It shows the proboscis of *C. gigas*, whereas the palpi are those of *C. leptorhynchus*. In other respects it resembles both species.	South of Australia at the fiftieth parallel.
1375 to 1600	*Colossendeis gigas*, n. sp. This seems to be the largest, not only of the species of the genus *Colossendeis*, but of all the different forms hitherto described.	Between Cape of Good Hope and Kerguelen; between Juan Fernandez and Valparaiso.
25	*Colossendeis kröyerii*, Wood-Mason (sp.). *Rhopalorhynchus kröyerii*, Wood-Mason. A small but very characteristic species of the genus *Colossendeis*.	Andaman Islands (Port Blair).
120	*Colossendeis robusta*, n. sp. A beautiful species, easily to be recognised by the form of its proboscis, &c.	Kerguelen.
1375 to 1600	*Colossendeis gracilis*, n. sp. The species *C. media*, *C. brevipes*, and *C. gracilis*, proposed by me, only show very slight differences. In regard to *C. gracilis*, I think it is a very interesting fact, that one of the specimens is furnished with distinct mandibles.	Between Cape of Good Hope and Kerguelen.
2225	*Colossendeis media*, n. sp. Characterised by the long claws of the legs, and by the peculiar shape of the last joints of the palpi.	West of Valparaiso.

Depth in Fathoms.	Name.	Geographical Distribution.
2650	*Colossendeis brevipes, n. sp. In all probability nearly allied to C. media, Hoek, and C. gracilis, Hoek. Perhaps not a distinct species but only a variety.	East of Buenos Ayres.
1250	*Colossendeis minuta, n. sp. A very slender and easily recognized species.	South of Halifax.
	Pasithoe, Goodsir (1842).	
	No mandibles; palpi eight-jointed; ovigerous legs nine- (nonne ten-) jointed.	
(?)	(?)Pasithoe vesiculosa, Goodsir, Edin. New Phil. Journal, vol. xxiii., 1842. Description and characterisation of species and genus insufficient.	Coast of England.
	Endeis, Philippi (1843).	
	No mandibles; palpi seven-jointed; ovigerous legs nine- (nonne ten-) jointed.	
(?)	Endeis didactyla, Philippi, Ueber die Neapolitanischen Pycnogoniden, Arch. f. Naturgesch, ix., 176, 1843. Body ovate, ovigerous legs present. I think this must be considered as a true representative of the genus Endeis, proposed by Philippi. Semper (Ueber Pycnogoniden, Arb. Zool. Zoot. Inst. in Würzburg, i. 281, 1874) considers it as a species of Pasithoe, Goodsir. But this question cannot be settled without more detailed descriptions of the forms in question.	Naples.
(?)	(?)Endeis gracilis, Philippi, Ueber die Neapolit. Pycnog., Arch. f. Naturg., ix. 176, 1843. This species has a very slender body. No ovigerous legs are mentioned: perhaps the eight-jointed palpi are in reality the ovigerous legs, and this is a species of Phoxichilus without palpi.	Naples.
	Discoarachne, Hoek (1880).	
	No mandibles; palpi five-jointed; ovigerous legs ten-jointed.	
Shore.	*Discoarachne brevipes, n. sp. A very curious species with a disciform body and short legs.	Near Cape Town.

Family III. PALLENIDÆ

Contains those Pycnogonida, which have strongly developed cheliform mandibles, and either rudimentary palpi or no palpi at all. Ovigerous legs are present in both sexes, and furnished with denticulate spines (*Pallene*), or present in both sexes and not furnished with denticulate spines (some species of *Phoxichilidium*), or present only in the males (other species of *Phoxichilidium*). Two genera: *Pallene*, *Phoxichilidium*.

Depth in Fathoms.	Name.	Geographical Distribution.
	Pallene, Johnston (1837).	
	Mandibles cheliform ; palpi wanting ; ovigerous legs ten-jointed, present in both sexes. The last four joints of the ovigerous legs often furnished with denticulate spines.	
Shore.	*Pallene brevirostris,* Johnston, An Attempt to ascertain &c., Mag. of Zool. and Bot., vol. i., 1837. *P. brevirostris,* Johnston; Grube, Mittheilungen über St Vaast la Hougue, &c., Verhandl. der Schl. Ges. f. vaterl. Cultur., 1859. *P. brevirostris,* Johnston, Hoèk, Ueber Pycnogoniden, Niederl. Arch. f. Zool., iii., 1877. A well-characterised species.	Northern Europe (Coast of England, France, of the Netherlands).
0 to 3	(?)*Pallene empusa,* Wilson, Pycnogonida of New England, Trans. Connect. Acad., vol. v. p. 9, 1880. This species is so closely allied to *P. brevirostris,* Johnston of the European coast, that it must be considered either as the same, or as a variety of that species.	Vineyard Sound, Noank, Connecticut (U. S. A.).
(?)	(?)*Pallene chelifera,* Claparède (sp.), Beobachtungen über wirbellose Thiere, p. 103, 1863. Claparède gives a very insufficient description of this species, which he considers as a *Phoxichilidium.* I think it most probable, that it is closely allied to *P. brevirostris,* Johnston.	French Coast.
(?)	*Pallene spinipes,* Fabr. *Pycnogonum spinipes,* Fabr. (sp.), Fauna Grönl., p. 232. *Pallene spinipes,* Fabr., Kröyer, Bidrag til Kundskab, Naturhist. Tidskr., Ny Raekke, i. 118, 1845. *Pallene spinipes,* Fabr., Jarzynsky, Præmissus Catalogus, Annales de la Soc. des Natur. de St Petersb., 1870.	Coast of South Greenland ; White Coast of Russian-Lapland.
38 to 40	**Pallene lœvis,* n. sp. A well-characterised *Pallene* species.	South Coast of Australia.
(?)	*Pallene chiragra,* Milne-Edw., Histoire Natur. des Crustacés, tom. iii. p. 535, 1840. The description of this species given by Milne-Edw., *loc. cit.,* is insufficient. Perhaps it is the same as the species for which I propose the name *P. australiensis.*	Australia (Jervis Bay).
38 to 120	**Pallene australiensis,* n. sp. Perhaps it is this species, which has been described by Milne-Edwards as *P. chiragra* (Hist. Nat. des Crust., iii. 535, 1840).	South-east Coast of Australia.
(?)	*Pallene grubii,* Hoek. I propose this name for the species described by Grube (Jahresb. der Schles. Ges. f. vaterl. Cultur., p. 54, 1869), and which Grube considered as a species of *Phoxichilidium.*	China Sea.
(?)	*Pallene longiceps,* Böhm, Sitzungsber. der Ges. naturf. Freunde zu Berlin, p. 59, 1879. A very curious species with rudimentary two-jointed palpi in the male sex.	Japan.
38	**Pallene languida,* n. sp. Nearly allied to *P. longiceps,* Böhm (Sitzungsber. der Ges. Naturf. Freunde in Berlin, 1869).	Melbourne (Australia).

Depth in Fathoms.	Name.	Geographical Distribution.
(?)	*Pallene intermedia*, Kröyer, Bidrag til Kundskab, Naturh. Tidskr., Ny Raekke, i. 119, 1845.	Coast of South Greenland.
(?)	*Pallene discoidea*, Kröyer, Bidrag til Kundskab, Naturhist. Tidskr., Ny Raekke, i. 120, 1845. *Pallene discoidea*, Kröyer, Jarzynsky, Pnemissus Catalogus, Annales de la Soc. des Natur. de St Petersbourg, 1870.	Coast of South Greenland, White Sea, Coast of Russian-Lapland, and North Norway.
12	*Pallene hispida*, Stimpson, Invertebrata of Grand Manan, p. 37, 1853. *Pseudopallene hispida*, Stimpson (sp.), Wilson, Amer. Jour. of Sc. and Arts, vol. xv. p. 200, 1878, Trans. Connect. Acad., v. 10, 1880. Wilson considers this species, which is a true *Pallene*, as representing a new genus, which he calls *Pseudopallene*. But there is no difference in the number of the joints of the ovigerous legs, and the presence or absence of auxiliary claws furnishes by no means a trustworthy ground for division of the genus. "This species is very similar to the last (*P. discoidea*, Kröyer) and a larger number of specimens may show them to be identical" (Wilson, *ibid.*).	Near Eastport, Maine ; off Grand Manan.
(?)	*Pallene lappa*, Böhm, Pycnogoniden des Museums zu Berlin, Monatsber. der K. Ak. der Wissensch. zu Berlin, 182, 1879. I was long in doubt whether this species was a true *Pallene* or a *Phoxichilidium*. I have arrived at the conclusion that it is indeed a *Pallene*, but a young one, with not quite developed ovigerous legs.	Mozambique.
(?)	(?)*Pallene circularis*, Goodsir, Edin. New Phil. Journal, vol. xxxii. p. 136, 1842. Goodsir's description is insufficient. Perhaps this species is nearly allied to *P. discoidea*, Kröyer, Bidrag til Kundskab, Naturhist. Tidskr., Ny Raekke, i. 120, 1845.	Scotland.
191 to 459	*Pallene malleolata*, G. O. Sars, Crustacea et Pycnogonida nova, Arch. f. Math. og. Naturvid., iv. 469, 1879. Seems to be a species characteristic of the higher northern latitudes.	Lat. 72° 27′ to 80° N., long. 5° 40′ to 20° 51′ E.

Phoxichilidium, Milne-Edwards (1840).

Basis of the proboscis dorsally covered by the front part of the cephalothoracic segment. On this front part the oculiferous tubercle is placed. Mandibles cheliform ; palpi wanting ; ovigerous legs five- to ten- (?) jointed, the last four joints never furnished with denticulate spines ; in some species present in both sexes, in others only in the male sex.

| Shore. | *Phoxichilidium femoratum*, Rathke (sp.). *Nymphon femoratum*, Rathke, Naturh. Selsk. Skr., v., i. 201, | Greenland, Russian - Lapland (Jarzynsky), Norway, Den- |

Depth in Fathoms.	Name.	Geographical Distribution.
	1799. *Orithyia coccinea*, Johnston, An Attempt, &c., Magazine of Zoology and Botany, vol. i. p. 378, 1837. *P. femoratum*, Rathke, Kröyer, Bidrag til Kundskab, Naturh. Tidskr. i., 1845 ; Hoek, Ueber Pycnogoniden, Niederl. Arch. f. Zool. iii., 1877. This is the largest of the *Phoxichilidium* occurring in the neighbourhood of the shores of the North Sea. (Ovigerous legs, occurring only in the males, five-jointed).	mark, Heligoland, England, Holland.
(?)	(?)*Phoxichilidium globosum*, Goodsir, Edin. New Philos. Journ., xxxii., 1842. Description quite insufficient, probably a female of some species or other.	North Atlantic (Orkney).
Low water mark (Wilson).	*Phoxichilidium maxillare*, Stimpson, Invertebrata of Grand Manan, 37, 1853. *P. maxillare*, Stimpson, Wilson, Pycnogonida of New England, Trans. Connect. Acad., vol. v. p. 12, 1880. I never saw this species. Wilson says it resembles *P. femoratum* of Europe. It has five-jointed ovigerous legs, wanting in the female.	Bay of Fundy, Casco Bay, Halifax (East Coast of North America).
Tide-pools.	*Phoxichilidium minor*, Wilson, Pycnogonida of New England, Trans. Connect. Acad., v. 13, 1880. Smaller than *P. maxillare*, to which it is closely allied. It has also five-jointed ovigerous legs, wanting in the female. It may be a dwarf variety of *P. maxillare* (Wilson).	Casco Bay (East Coast of North America).
(?)	(?)*Phoxichilidium mutilatum*, Frey u. Leuckart, Beiträge zur Kenntniss wirbelloser Thiere, p. 165, 1847. With the fourth pair of legs rudimentary and one-jointed ovigerous legs. No doubt the description is taken from an immature specimen.	Heligoland.
30 to 42 (Böhm). 7 to 20 (Challenger).	**Phoxichilidium fluminense*, Kröyer, Bidrag til Kundskab, Naturh. Tidskr., N. R., i., 1845. *Pallene fluminensis*, Kröyer (sp.), Böhm, Pycnogoniden des Museums zu Berlin, Monatsber. der Königl. Akad. der Wiss., p. 180, 1869. Dredged during the Challenger Expedition off Bahia. Ovigerous legs in both sexes, ten-jointed.	Coast of Patagonia, Magellan Strait, Coast of Brazil.
7 to 20	**Phoxichilidium insigne*, n. sp. A very characteristic species, with an extremely slender body armed with spine-forming knobs.	Bahia.
(?) (Probably 25 to 35.)	*Phoxichilidium petiolatum*, Kröyer, Bidrag til Kundskab, Naturh. Tidskr., i., 1845. *Pallene attenuata*, Hodge, Report British Association, 119, 1862. After Hodge (Ann. and Mag. of Nat. Hist., vol. xiii, 3d series, p. 116, 1864). *Pallene attenuata* is a synonym of Kröyer's *P. petiolatum*. Ovigerous legs seven-jointed, wanting in the female.	Coast of Norway (Oeresund), Coast of Northumberland, Doggerbank.

Depth in Fathoms.	Name.	Geographical Distribution.
48 to 175	*Phoxichilidium patagonicum, n. sp. A beautiful species. It was dredged at different stations in the neighbourhood of the Patagonian coast.	Patagonia.
600	*Phoxichilidium patagonicum, var. elegans, Hoek. Whether this is a true variety of the preceding species, or is only to be considered as a young specimen of this species, I am unable to ascertain.	East of Cape Corrientes (La Plata).
(?)	Phoxichilidium digitatum, Böhm, Pycnogoniden des Museums zu Berlin, Monatsber. der K. Akad. der Wiss. in Berlin, 184, 1879. Perhaps nearly allied to my P. patagonicum (ovigerous legs occurring only in the males. Number of joints?).	Singapore.
(?)	(?)Phoxichilidium pygmæum, Hodge (sp.). Pallene pygmæa, Hodge, British Pycnogonoidea, Ann. and Mag. of Nat. Hist., 3d series, vol. xiii., 1864. As far as I could ascertain from the extremely insufficient description of Hodge, and from his figures, this species belongs to the genus Phoxichilidium. One of the species of Phoxichilidium occurring on the French and on the Dutch coast is probably the same species. Should this prove to be true the ovigerous leg is six-jointed and occurs only in the male.	England.
Shore.	Phoxichilidium virescens, Hodge, British Pycnogonoidea, Ann. and Mag. of Nat. Hist., 3d series, vol. xiii., 1864. Description insufficient. As far as I could make out it is one of the more common forms of the Dutch and French coast. (Ovigerous legs wanting in the female, six-jointed.)	English Coast (Polperro), French Coast (Grube, Mittheilungen über St Malo u. Roscoff).
Tide-marks to 6 fathoms.	Phoxichilidium lentum, Wilson (sp.). Anoplodactylus lentus, Wilson, Pycnogonida of New England, Trans. Connect. Acad. of Arts and Sc., vol. v. p. 14, 1880. Wilson gives this name to the species described by Smith (Report Invertebrata of Vineyard Sound, p. 250) as Phoxichilidium maxillare. I think it is closely allied to Phoxichilidium virescens, Hodge (Ann. and Mag., 3d. ser., vol. xiii. p. 115, 1864). Ovigerous legs only in the male, six-jointed.	Vineyard Sound; Bay of Fundy (New England).
1675	*Phoxichilidium oscitans, n. sp. A very curious deep-sea species, easily to be recognised.	Atlantic : West of Azores.
1600 to 1950	*Phoxichilidium pilosum, n. sp. A beautiful deep-sea species.	Between Cape Town and Kerguelen; between Kerguelen and Melbourne.
1875	*Phoxichilidium mollissimum, n. sp. Unfortunately only a defective specimen of this curious form was dredged by the Challenger Expedition. It has ten-jointed ovigerous legs.	Off Yeddo (Japan).

Depth in Fathoms.	Name.	Geographical Distribution.
(?)	*Phoxichilidium johnstonianum,* White (sp.). *Nymphon johnstonianum,* White, Proc. Zool. Soc. of London, vol. xv., 1847. There can be no doubt that this species belongs to *Phoxichilidium :* eyes situated above the insertion of the chelicera (mandibles); beak (proboscis) springing from the under side of the head; chelicera with two basal joints, &c. The description, however, is hardly sufficient. What White describes as palpi are in all probability the ovigerous legs.	South Seas.
(?)	(?)*Phoxichilidium phasma,* White (sp.). *Nymphon phasma,* White, Proc. Zool. Soc. of London, vol. xv., 1847. Whether this species also belongs to this genus or is to be considered as a *Pallene* (it is certainly not a *Nymphon*) is not to be ascertained. White says it may possibly be the other sex of the preceding	South Seas.
	Oomerus, Hesse (1874).	
	Oculiferous tubercle placed at the base of the proboscis; mandibles with long pincers; palpi represented by small knobs; ovigerous legs not present in the female.	
(?)	(?)*Oomerus stigmatophorus,* Hesse, Ann. d. Sc. Nat. Zool., 5ième série, xx., 1874, art. 5, p. 18, pl. viii. In all probability this is a species of the genus *Phoxichilidium,* Milne-Edwards. Only a female without ovigerous legs, and with highly developed ovaries in the fourth joint of the leg, was observed by Hesse.	Brest (Bretagne).

Family IV. Phoxichilidæ

Contains those Pycnogonida, which have neither mandibles nor palpi, or have them rudimentary. Ovigerous legs, as a rule, only in the males, whereas *Hannonia* possesses them in both sexes; always without denticulate spines. Genera: *Hannonia, Phoxichilus, Pycnogonum.*

Depth in Fathoms.	Name.	Geographical Distribution.
	Hannonia, Hoek (1880).	
	Mandibles rudimentary, chelate, two-jointed; no palpi; ovigerous legs ten-jointed, present in both sexes.	
Shore.	*Hannonia typica,* n. sp. A short-legged species, a true littoral form, with very characteristic features.	Cape of Good Hope.

Depth in Fathoms.	Name.	Geographical Distribution.
	Phoxichilus, Latr. (1816).	
	Mandibles and palpi wanting; ovigerous legs, seven-jointed, present only in the males; body, as a rule, slender.	
Shore.	*Phoxichilus spinosus*, Montagu (sp.). *Phalangium spinosum*, Montagu, Linn. Transact., vol. ix. p. 100, pl. v. fig. 7, 1808. *P. spinosus*, Montagu, Johnston, An Attempt, &c., Magaz. of Zool. and Bot., i. 1837. *Phoxichilus spinosus*, Mont., Kröyer, Bidrag til Kundskab Naturh., Tidskr. Ny. Raekke, i., 1845. The *Phoxichilus spinosus*, Leach, as described and figured by Quatrefages (Ann. d. Sc. Nat., 3ième série, Zool., tom. iv. 1845) is not a *Phoxichilus* at all. Perhaps it is a *Pallene*.	Coast of Norway; Coast of Russian-Lapland (Jarzynsky); England (South Coast of Devonshire).
(?)	*Phoxichilus meridionalis*, Böhm, Pycnogoniden des Museums zu Berlin, Monatsber. der Königl. Akad. der Wissensch. zu Berlin, 1879, s. 189, Taf. ii., fig. 4–4b. A species which stands between *P. lævis*, Grube, and *P. inermis*, Hesse (Böhm).	Singapore.
(?)	(?)*Phoxichilus inermis*, Hesse, Ann. d. Sc. Natur., 5ième série, tom. vii. p. 199, 1867. Perhaps nearly allied to *P. lævis*, Grube. Judging from Hesse's description, this species is furnished with a three-jointed abdomen, and this is not the case with Grube's species, nor perhaps with any species of the Pycnogonida. It was taken at Brest on the keel of a ship returning from the Mediterranean.	French Coast or Mediterranean (?).
Shore.	*Phoxichilus lævis*, Grube, Mittheilungen über St Malo und Roscoff, 1872. During a recent visit to Roscoff (Bretagne) I collected specimens of this species, which is easily distinguished from *P. spinosus*, Montagu.	French Coast.
(?)	(?)*Phoxichilus pigmæus*, Costa (*Foxichilus pigmæus*), Fauna del Regno di Napoli, Napoli, 1836, p. 10. This species, characterised by its smallness and by the absence of the spines at the end of the joints of the legs, is perhaps identical with *P. inermis*, Hesse, or *P. lævis*, Grube. Description very insufficient.	Gulf of Naples.
	Pycnogonum, Brünnich (1764).	
	Mandibles and palpi wanting; ovigerous legs nine-jointed, present only in the males; body, as a rule, robust.	
Tide marks to 430 fathoms.	*Pycnogonum litorale*, Ström (sp.). *Phalangium litorale*, Ström, Physisk. og œconomisk beskrivelse over fogderiet Söndmör, Sorüe, 1762. *P. litorale*, Ström, Kröyer, Bidrag til Kundskab, Naturh., Tidskr. Ny Raekke, i. 126, 1845. *P. litorale*, O. Fabr.,	North European Coasts and Seas; Coast of Russian Lapland (Jarzynsky); Coast of North America, as far south as Long Island Sound; Coast of

Depth in Fathoms.	Name.	Geographical Distribution.
	Wilson, Pycnogonida of New England, Transact. Connect. Acad., vol. v. p. 4, 1880. In the Ann. and Mag. of Nat. Hist., 5th series, vol. iii. p. 283, 1879, Slater describes a variety of this species, which, being very slender, is named *P. litorale*, var. *tenue*, Slater. It was dredged by Capt. St John and placed in the British Mus. Catal., 78, 11.	Chili (Gay); Mediterranean; Japan. The greatest depth from which it has been obtained in Europe is, as far as I could ascertain, 53 fathoms (" Knight Errant "), whereas it was dredged by Smith and Harger (1872) at a depth of 430 fathoms, east of St George's Bank, N. lat. 41° 25', long. 65° 42' 3" W.
(?)	(1)*Pycnogonum australe*, Grube, Jahresb. der Schles. Ges. f. vaterl. Cultur, p. 54, 1869. Of this species only a larva with three pairs of legs has been examined. It seems to be a species with auxiliary claws at the ends of the legs.	Australia.

DESCRIPTION OF THE SPECIES DREDGED DURING THE CHALLENGER EXPEDITION.

Nymphon, Fabr.

Nymphon hamatum, n. sp. (Pl. I.).

Diagnosis.—Body slender, body and legs almost entirely smooth; eyes obsolete, auxiliary claws wanting; second joint of the palpi longer than the third; second joint of the leg longer than the first and the third; second tarsal joint of the leg longer than the first.

Description.—The body is slender, and the lateral processes are separated. The proboscis is large, almost one-third the length of the body, slightly swollen in the middle, and again at the extremity. The mouth is triangular, not very large. The cephalothoracic segment (with the base of the mandibles swollen) is almost as long as the proboscis. The eyes are obsolete, represented only by two small knobs behind the lateral process of the cephalothoracic segment. The abdomen is rather large; the mandibles large, with the basal joint as long as the rostrum; the claws of the chelæ are elongated; the immovable claw more strongly curved than the movable one (Pl. I. fig. 3). Both claws are armed with spines; on the movable claw they are more numerous and larger (fig. 3). Seen but slightly magnified, the mandibles are smooth; when greatly magnified they show small hairs all over the surface.

The palpi are slender, longer than the rostrum; the second joint is longer than the third, the fifth longer than the fourth; they increase in length as follows:—First, fourth, fifth, third, second. The first and second joints are almost entirely smooth, the third joint with small and the fourth and fifth with stronger hairs.

The ovigerous legs of the males are stronger than those of the females, and in all the specimens are bent as shown in the figure (fig. 2). The fourth joint is curved, the fifth thinner, and much longer than the fourth, and swollen at the extremity ; the sixth is short, the seventh, eighth, ninth, and tenth very short ; the first joints are sparsely hairy, the fifth not very hairy, the sixth hairy, with a row of stronger hairs at the outer extremity of the joint. The spines of the four last joints are not very denticulated (figs. 4 and 5), their numbers are respectively 12, 10, 9, 12 ; the end claw is denticulated also. ·

In the females the ovigerous legs are shorter, and not bent as in the males. The fifth joint is only a little longer than the fourth, the sixth joint is less hairy, the denticulated spines of the four last joints not so numerous, their numbers being respectively 11, 7, 5, 7.

The legs are very long, measuring, for instance, 38 mm. in a female of 11 mm., and 44 mm. in a male of 13·5 mm. (1 : 3·4, and 1 : 3·3); the second joint, which is swollen at the extremity in the females is longer than the first and third joints ; the fifth joint is the longest, the sixth not much shorter ; the second tarsal joint is longer than the first, the claw is not very strong, nearly half the length of the second tarsal joint (fig. 9). Auxiliary claws are wanting. The fourth joint of the leg, which in the females is swollen with the ovary, is furnished at the extremity with a hook-like process bearing one or two hairs (fig. 8). I believe this is the first species of *Nymphon*, in which this process has been observed, and therefore I have named the species after it. This fourth segment is furnished in the males with a row of knobs, closed at the extremity by a thin perforated membrane (fig. 7). Both males and females have the legs almost entirely smooth, the hairs being so small as to be only visible under the microscope. Larger hairs are seen at the extremity of the joints. The last joints are furnished with small but very dense hairs. The genital openings of the females are large, and easily observed on the lower side of the second joint on each leg (fig. 6). Those of the males are a great deal smaller, and six in number ; they are not found at the first pair of legs. The colour of alcoholic specimens is light yellowish (for the larvæ see below).

Habitat.—This very beautiful species was dredged during the Challenger Expedition between the Cape of Good Hope and Kerguelen, off the Crozets Islands. There are in all eight specimens, of which four are males and four females. One of the males was furnished with eggs, or rather with young ones, adhering still to the accessory legs. The species was found at two stations, at 1375 and 1600 fathoms. At the same time were obtained two specimens of *Ascorhynchus glaber*, Hock, two of *Colossendeis gigas*, Hock, three of *Colossendeis leptorhynchus*, Hock, three of *Colossendeis gracilis*, Hock, and one of *Phoxichilidium pilosum*, Hock.

Station 146.   December 29, 1873.   Lat. 46° 46′ S., long. 45° 31′ E.   Depth, 1375 fathoms.   Bottom temperature, 1·5° C.   Sea bottom, globigerina ooze.

Station 147.   December 30, 1873.   Lat. 46° 16′ S., long. 48° 27′ E.   Depth, 1600 fathoms.   Bottom temperature, 0·8° C.   Sea bottom, globigerina ooze.

Observations.—*Nymphon hamatum* is a very fine deep-sea Pycnogonid, and may easily be distinguished from the other species.   Among the described species of *Nymphon* it shows some resemblance to *Nymphon macronyx*, Sars, but this species is a great deal smaller, has the mandibles and the legs shorter, shows a very prominent and curious-shaped oculiferous tubercle,[1] and has the claw of the leg as long as the second tarsal joint.

Nymphon longicoxa, n. sp. (Pl. II. figs. 1–5 ; Pl. XV. figs. 8, 9).

Diagnosis.—Body very slender and smooth ; legs almost entirely smooth ; eyes small but distinct, oculiferous tubercle rounded ; auxiliary claws wanting ; second joint of the palpi very long, much longer than the third ; second joint of the feet much longer than the first and the third, the sixth joint the longest, the second tarsal joint longer than the first.

Description.—The body is very slender, the lateral processes with large intervals between them.   The proboscis is large, one-third of the length of the body, in general resembling that of *Nymphon hamatum*, but a little narrower.   The mouth is triangular, not very large.   The cephalothoracic segment is as in *Nymphon hamatum*.   The eyes are rudimentary, four, situated on a rounded tubercle.   The abdomen is longer than in *Nymphon hamatum*.

The mandibles are very long, the basal joint longer than the rostrum, the second joint also very long.   The immovable claw, which is curved more strongly than the movable one, is furnished with very large spines, which reach almost to the extremity (Pl. II. fig. 3).   The movable claw furnished with smaller spines has the extremity smooth ; the mandibles are smooth, the second joint only furnished with microscopic hairs.   The palpi are extremely slender, longer than the rostrum, the second joint is very large, the fourth and fifth almost equal, the latter furnished with small hairs (fig. 2).

The ovigerous legs of the full-grown males are characteristic.   The fifth joint is very long, and describes an elegant curve ; it is divided into two parts by a rudimentary articulation, and is strongly swollen at the extremity.   The sixth joint, which is also curved, makes an angle with the foregoing.   The four last joints are small, and often bent so as to describe a spiral.   The first joints are smooth, at the end of the fifth there is, on the outside, a small quantity of hairs, the sixth is furnished with numerous hairs, and has on the upper surface rows of knobs of a curious shape.   I have figured some of them (Pl. XV. fig. 8).   They are also present on the fifth joint, but are smaller and not so numerous.   The spines of the four last joints are much denticulated (Pl. II. fig. 4) ; their numbers are respectively 13, 8, 7, 6.   The spines of the end-claw are very small and blunt.

[1] See the description hereafter in the Appendix.

The ovigerous leg of the full-grown female is almost entirely straight. The difference in length between the fourth and the fifth joints is not so considerable as in the males; the denticulated spines on the four last joints are more numerous than in the males, their numbers being 19, 12, 10, 9. These curiously-shaped knobs do not occur on the ovigerous leg of the female.

The legs are still longer than those of *Nymphon hamatum*. In a female of 12 mm. they measured 46 mm.; in a male of 9½, 38 mm. (1 : 3·8 and 1 : 4). The second joint is in the males four times as long as the first, in the females a little shorter, but considerably swollen at the extremity; the sixth joint is the longest, being more than once and a half the length of the fifth; the first tarsal joint is shorter than the second, the claw is almost as long as the first tarsal joint, auxiliary claws are wanting (Pl. II. fig. 5). The first joints of the legs are almost entirely smooth, the hairs increasing in number as they approach the extremity of the leg. The genital openings of the females are very large, and are found on every leg. Those of the males are smaller, and found only on the three hinder pairs of legs (Pl. XV. fig. 9). The colour of alcoholic specimens is light yellowish. (For the larvæ see below.)

Habitat.—This fine species was dredged east of Auckland. There are in all twelve specimens, of which only three are females. One of the males was furnished with larvæ clinging to the accessory legs. The depth at which the specimens were found is 1100 fathoms. At the same place two specimens of *Nymphon compactum*, Hoek, were obtained.

Station 168. July 8, 1874. Lat. 40° 28′ S., long. 177° 43′ E. Depth, 1100 fathoms. Bottom temperature, 2·0° C. Sea bottom, grey ooze.

Observations.—I believe this species with its rudimentary eyes to form the transition· from the shallow-water species to the true deep-sea species. The very long coxæ render the species easily distinguishable.

Nymphon procerum, n. sp. (Pl. II. figs. 9–12).

Diagnosis.—Body extremely slender, smooth; legs hairy; eyes obsolete; auxiliary claws wanting; the second joint of the palpi a little longer than the third, the second joint of the leg longer than the first and the third, the second tarsal joint of the leg a little longer than the first.

Description.—The body is very slender, and the lateral processes are separated by large intervals. The proboscis is slender, shorter than one-third of the length of the body, in the middle a little thicker. The cephalothoracic segment is longer than the proboscis. Eyes are wanting; the abdomen is small and bent upwards. The mandibles are very long, the basal joint longer than the rostrum, the second joint also long and slender, the claws very long; the spines of the movable claw are smaller, and closer to one another than are those of the immovable one (Pl. II. fig. 10). The palpi are very slender, much longer than the rostrum; the second joint is a little longer than the third; the fourth and fifth together are as long, or a little longer than the second joint. The palpi are nearly hairless, only the last joints being furnished with very small hairs. The ovigerous legs are feeble, shorter than the length of the body, the fourth joint has a distinct knob at a distance of nearly a third of its length, measured from the beginning; the fifth joint is the longest, the sixth half the length of the fifth, the seventh to the tenth armed with sharply denticulated spines (Pl. II. fig. 12); the claw has numerous and dense spines (Pl. II. fig. 11). The ovigerous legs are almost entirely smooth.

The legs are slender, being more than three times as long as the very long and slender body (body 12 mm., legs 38 mm.). The second joint is longer than the first and third, and is considerably swollen; the fourth joint is swollen with the ovaries, and is nearly as long as the fifth joint, the sixth joint is the longest. The two tarsal joints describe a slight curve, the second is a little longer than the first, the claw is short, auxiliary claws are wanting. The fourth, fifth, and sixth joints have long but not very dense hairs; the seventh and eighth joints have denser but very small hairs. The second joint of each leg is furnished with a large genital opening.

Habitat.—The single female specimen of this species was dredged West of Valparaiso, at a depth of 2160 fathoms.

Station 299. December 14, 1875. Lat. 33° 31′ S., long. 74° 43′ W. Depth; 2160 fathoms. Bottom temperature, 1·1° C. Sea bottom, grey mud.

Observations.—There can be little doubt, I believe, that this species is closely allied to the two foregoing species. Yet I think its extremely slender and elongated form of body characteristic enough to establish a new species upon it. With the exception of *Nymphon longicollum*, dredged from a depth of 2225 fathoms, of all the genus *Nymphon procerum* inhabits the greatest depth.

Nymphon longicollum, n. sp. (Pl. III. figs. 1–3; Pl. XV. fig. 11).

Diagnosis.—Body slender; distance between the insertion of the rostrum and the attachment of the ovigerous legs very great; eyes obsolete, auxiliary claws extremely small. The second joint of the palpi twice as long as the third, the second joint of the leg three times as long as the first, the second tarsal joint of the leg nearly as long as the first.

Description.—The body is slender, the proboscis long, and exactly cylindrical; the cephalothoracic segment is much longer than one-third of the length of the body. The intervals between the lateral processes of the body are very large, they are totally wanting between the attachment of the ovigerous leg and of the first true leg. Eyes are wanting, but the conical oculiferous tubercle is very large and acute. The abdomen is small.

The mandibles have the basal joint as long as the rostrum, the second joint is short, and furnished with short claws. The movable claw is a little longer than the immovable one, the former is curved at the extremity, the latter straight. The spines on the claws are not very prominent (Pl. III fig. 3). The two joints are sparsely hairy, but the hairs of the second joint are longer than those of the first.

The palpi are not very long, and are feeble. The second joint is considerably longer than the third ; the first and second together are nearly as long as the last three together. The hairs are much more numerous on the outer joints than on the first two.

The ovigerous legs are not very long. The first four joints are almost entirely smooth, with the exception of some long hairs at the extremity of the fourth joint ; the fifth and sixth are hairy when seen through the microscope, the spines of the four last joints are sharply denticulated, their numbers being respectively 9, 6, 5, 5. On the claw there are five not very strong spines. The fifth joint of the ovigerous leg is the longest.

The legs are very slender. The length of the body of the single specimen is nearly 6 mm., that of the leg 26 mm. The second joint is three times as long as the third, the fourth and fifth are nearly equal, the sixth united with the two tarsal joints are as long as the fourth and fifth together. The first tarsal joint is at the first leg a great deal shorter than that of the second. In the other legs the difference between the two tarsal joints is not so considerable ; the claw is half as long as the second tarsal joint. The auxiliary claws are extremely small (Pl. XV. fig. 11). The legs, when examined with a magnifying glass of small strength are quite smooth ; when magnified greatly they show small hairs, which increase in number and size towards the extremity of the leg (?). The single specimen of this species brought home by the Challenger is in all probability a male. The animal is perhaps a young one, as I failed to observe the genital openings.

Habitat.—The specimen was dredged off the coast of Chili.

Station 298. November 17, 1875. Lat. 34° 7′ S., long. 73° 56′ W. Depth, 2225 fathoms. Bottom temperature, 1·3° C. Sea bottom, grey mud.

Observations.—This curious species is very easily distinguished by its extremely long neck and legs, the latter being more than four times as long as the body. From its long slender neck it bears a certain resemblance to *Nymphon longitarse*, Kr. It is a true deep-sea species.

Nymphon compactum, n. sp. (Pl. II. figs. 6–8 ; Pl. XV. fig. 10).

Diagnosis.—Body stout, sparsely hairy ; eyes obsolete ; auxiliary claws wanting. Second joint of the palpi longer than the third, the second joint of the leg longer than the first, the second tarsal joint shorter than the first.

Description.—The body is stout, the proboscis thick and swollen a little in the middle, and again at the extremity ; the length about one-third of the length of the body. The cephalothoracic segment is short, swollen anteriorly, and constricted in the middle. Eyes are wanting, the oculiferous tubercle is represented by a blunt knob (fig. 7). The abdomen is long. The intervals between the lateral processes of the body are small. The body is almost smooth, and the lateral processes are furnished with long hairs. The mandibles are long, the first joint a little curved, and much longer than the proboscis; at the ventral side this joint shows feebly an articulation near the base; the second is shorter, but furnished with very long claws, which are curved at the extremity, and both are armed with almost the same number of spines. The hairs on the mandibles are distant, the largest quantity being observed on the base of the immovable claw. The palpi are not very slender, yet a great deal longer than the proboscis; the second joint is the longest, then follow the third, the fifth, the fourth, and the first, which is the shortest of all. On the first two joints the hairs are not so numerous as on the last three joints.

The ovigerous legs of the female (the two specimens dredged by the Challenger were both females) are tolerably long and stout, once and a third as long as the body; the fourth, fifth, and sixth joints are nearly of the same length, the fourth being the longest, and the sixth the shortest. The four last joints again are of about the same length, the claw is slender, its length two-thirds the length of the last joint. Joints one to four are almost entirely smooth, with the exception of a row of hairs at the end of the fourth joint; joints five and six are furnished with numerous spines, the last four joints showing hairs only at the distal extremity. The spines of the four last joints (figured on Pl. II. fig. 8) are of a very irregular shape, and not very numerous, their numbers being respectively 10, 8, 5, 7. The spines of the claw are not very strong.

The legs are three times as long as the body (36 mm. in a body of 12 mm.). The second joint is longer than the first and third, and swollen, as is always the case in the females. The fourth joint is also very considerably swollen (with the ovaries), the fifth and the sixth joints are nearly of the same length, and a little longer than the fourth; of the two tarsal joints, which are together nearly two-thirds the length of the sixth joint, the first is longer than the second (Pl. XV. fig. 10). The claw is half the length of the second tarsal joint, auxiliary claws are wanting. The fourth joint of the leg is sparsely hairy, the fifth a little more so, the sixth is very hairy and shows some strong spines at the extremity, the two tarsal joints are covered with very minute hairs. The genital openings are large and easily seen.

Habitat.—Of this species two females were dredged along with *Nymphon longicoxa*, east of Auckland.

Station 168. July 8, 1874. Lat. 40° 28′ S., long. 177° 43′ E. Depth, 1100 fathoms. Bottom temperature, 2·0° C. Sea bottom, grey ooze.

Observations.—*Nymphon longicoxa* and *Nymphon compactum* were obtained from a depth of 1100 fathoms. *Nymphon longicoxa* shows rudimentary eyes, those of *Nymphon compactum* are quite obsolete. *Nymphon longicoxa* is one of the most slender, *Nymphon compactum* one of the stoutest species dredged by the Challenger. In the one the auxiliary claws are wanting, whereas small ones are present in *Nymphon longicoxa*, and in every other respect they are as widely different as two species of the same genus of Pycnogonids can be. *Nymphon compactum* shows some relationship to *Nymphon strœmii*,

Kr., but my species may be readily distinguished by its very short cephalic segment, the absence of auxiliary claws, of eyes, &c.

Nymphon meridionale, n. sp. (Pl. III. figs. 4–8).

Diagnosis.—Body slender and smooth, legs not very hairy, slender; eyes four, distinct; auxiliary claws present; second joint of the palpi longer than the third; second joint of the legs elongated, second tarsal joint of the leg shorter than the first.

Description.—The body of this species is slender, the lateral processes are widely separated with the exception of the lateral process of the first leg, and the small process of the ovigerous leg. The proboscis is cylindrical, its length is almost one-third of the length of the body. The cephalothoracic segment with the base of the mandibles is considerably swollen, very large, longer than the two following segments united. The eyes are distinct, four, placed round a small conical tubercle. The abdomen is small.

The mandibles are very long and robust. The first joint is a little curved, and longer than the rostrum, the second joint is also large, and furnished with very strong claws (Pl. III. fig. 5). The movable claw is longer and more deeply curved than the immovable one; both are furnished with a row of blunt and strong spines. The immovable claw shows numerous hairs which are also observed at the base of the movable one. The palpi are very long, the second joint is the longest, the third reaching farther than the rostrum, the fourth and the fifth furnished with numerous hairs, and about the same length.

The ovigerous legs are comparatively small in the single specimen dredged by the Challenger, which I think is a male not yet fully grown. The fifth joint is the longest; the four last joints are furnished with very numerous denticulated spines, their numbers being respectively 17, 16, 13, 13. These spines are comparatively small; they are elongated, and densely denticulated (Pl. III. fig. 6). The claw is not very long, and is furnished to the end with a row of short and blunt spines. All the joints of the ovigerous legs are smooth, with the exception of a few hairs placed at the extremity of the fourth and tenth joints.

The legs are long and slender; the specimen of 6½ mm. shows legs of 21½ mm. The second joint is more than twice as long as the third; the sixth joint is the longest, the second tarsal joint is shorter than the first (Pl. III. fig. 7), the claw is very small, the accessory claws are half as long as the claw (fig. 8); longer hairs are placed at the extremity of every joint, shorter ones cover the outer joints all over, and are a great deal less numerous on the inner joints. The genital openings I could not observe.

Habitat.—The only specimen was dredged in the Antarctic Ocean, at Station 153. February 14, 1874. Lat. 65° 42′ S., long. 79° 49′ E. Depth of the sea, 1675 fathoms. Sea bottom, mud.

Observations.—This is the most southern species of *Nymphon* (of Pycnogonids in general) hitherto observed. I think this form is closely related to *Nymphon gracilipes*, from Kerguelen ; yet the two species may easily be distinguished from each other by the form of the oculiferous tubercle, by the length of the claws of the mandibles, and by the length of the legs, which are in *Nymphon gracilipes*, Miers (after Böhm), five times as long as the body. It is remarkable that this species living at a depth of 1675 fathoms should have normally developed eyes.

Nymphon grossipes, Oth. Fabr. sp., (Pl. III. figs. 9–12 ; Pl. IV. fig. 1).

    Pycnogonum grossipes, Oth. Fabr., Fauna Groenlandica, p. 229, 1780. ·

    Nymphon grossipes, Oth. Fabr., Kröyer, Bidrag til Kundskab, Naturh. Tidskr. N. R., vol. i.
        p. 108, 1845.

    Nymphon grossipes, Oth. Fabr., Wilson, Pycnogonida of New England, Transactions Con-
        necticut Acad., vol. v. p. 21, 1880.

Description.—The body of this species is slender and almost smooth, the lateral processes are widely separated, with the exception of the small lateral processes of the ovigerous legs, and those of the first pair of true legs, between which no interval is observed. The proboscis is not very long, cylindrical, a little swollen at the extremity. The cephalothoracic segment is longer than the rostrum, swollen considerably at the base of the mandibles. The abdomen is small. The oculiferous tubercle is very prominent, conical, acute. The eyes are four, large (Pl. IV. fig. 1).

The mandibles have a long basal joint, which is longer than the proboscis and narrower ; the second joint is not very long; the claws are short, not very hairy, but armed with numerous equi-distant spines (Pl. III. fig. 10). The two mandibles are, in the three specimens procured, strongly divergent. The palpi are not very slender, nearly one-half as long as the proboscis, with the second joint not quite as long as the third, and the fifth longer than the fourth, furnished with numerous hairs at the end of the third joint, on the fourth, and on the fifth joint.

The ovigerous legs of the males (Pl. III. fig. 9) are long, more than one-half longer than the body, the fourth joint is the longest, the fifth nearly as long, all the joints are covered with very small perpendicular hairs. The spines of the four last joints are small, but numerous and elongated, sharply serrated (Pl. III. fig. 11). The claw is small, with numerous thin spines.

The ovigerous legs of the female are much shorter; 8 mm. in a female of 7 mm. The relative length of the joints is the same as in the male. The hairs are much smaller. The numbers of the denticulated spines on the four last joints are respectively 18, 17, 16, 14. There are about sixteen very slender and pointed spines at the claw.

The legs are long and slender, nearly five times as long as the body ; a male of 8 mm. has legs of 38 mm. The joints are sparsely hairy, with a row of stronger hairs at the junction of two joints ; the second joint is twice as long as the first ; the sixth joint is by far the longest. The first tarsal joint is longer than the second, which is armed with

---

[1] The figures on Plates III. and IV. belonging to this species are marked, *N. armatum*, n. sp.

a row of very strong hairs placed between the thinner ones (fig. 12). The auxiliary claws are longer than half the length of the claw.

The one female specimen shows large genital openings on the second joint of every leg; the two males have the openings a great deal smaller, and only on the six hind legs. The female is immature. Its length is 7 mm., that of the males 8 mm.

Habitat.—Of this species three specimens were dredged along with *Nymphon brevi-collum*, south of Halifax.

Station 49. May 20, 1873. Lat. 43° 3′ N., long. 63° 39′ W. Depth, 83 fathoms. Bottom temperature, 1·6° C. Sea bottom, gravel, stones.

Observations.—The specimens of this species have long been considered by me as closely related to, yet distinct from the *Nymphon grossipes*, O. Fabr. When I first examined this species, and made the drawings as figured on Plate III., I knew *Nymphon grossipes* only by the description of Kröyer. Since that time Mr Wilson's paper was published (March 1880), and as soon as I read his description of *Nymphon grossipes*, I almost felt sure that he had examined specimens quite identical with mine, and had rightly considered them as *Nymphon grossipes*. Lately I have had an opportunity of comparing the specimens of *Nymphon grossipes* dredged during the Challenger cruise with others, undoubtedly belonging to the same species, collected during the two cruises of the Dutch schooner "Willem Barents" in the Barents Sea. As there are however a few differences to be pointed out, it would perhaps be better to consider the Challenger specimens as forming a variety of this species, yet I hesitate to make such a proposal, as it is impossible to settle these questions without large collections from the same, or about the same, localities.

Nymphon brevicollum, n. sp. (Pl. III. figs. 13–15, Pl. XV. figs. 12, 13).

Diagnosis.—Body not very slender; legs slender, smooth; neck short; a small interval between the attachment of the ovigerous leg and the first true leg; intervals between the lateral processes short; oculiferous tubercle blunt; second joint of palpi longer than third, second joint of the legs longer than first and third, second tarsal joint shorter than the first. Auxiliary claws present.

Description.—The proboscis of this species is comparatively large and nearly cylindrical. The neck is short, yet the segment formed by the conjunction of the cephalic and the first thoracic segment is tolerably large, there being also an interval (longer in the females than in the males) between the ovigerous leg and the first true leg. The abdomen is small. The oculiferous tubercle rounded and small, with four brown eyes.

The mandibles vary considerably; they often have the first joint shorter than the rostrum, the second joint comparatively long, the claws long also, armed with numerous small spines. Those of the males are more slender than those of the females.

The palpi are very slender; the first and second joint forming in the females a right angle with the proboscis. The second joint is very long, much longer than the third. The fourth and fifth are nearly equal, shorter than the third joint, together nearly as long as the second. The hairs on the first three joints are few, on the last two numerous, closely adhering to the joints. The palpi of the males show the same length relatively to the joints, but they are more parallel with the direction of the proboscis.

The ovigerous legs are slender, only a little longer than the body in the female. The fourth and fifth joints are nearly of the same length, the sixth shorter. The four last joints are strongly bent, and describe a spiral. The denticulated spines are long and slender, their numbers being respectively 17, 14, 11, 13. The claw is very large, and furnished with numerous small spines (Pl. III. fig. 14). The first four joints are almost smooth, the number of hairs increasing from the fifth to the tenth joint, specially large hairs being placed at the end of the joints.

In the males the ovigerous legs are a great deal longer, nearly 11 mm. when the body is 6 mm.; the fourth joint is curved; the fifth forms an angle with the fourth, is very long, and shows a rudimentary articulation (Pl. XV. figs. 12, 13) at two-thirds of the length of the joint from the beginning; the sixth joint is short, but swollen; the four last joints and claw as in the female, the denticulated spines being more numerous, and respectively 19, 16, 12, 15.

The legs of the females are more than four times as long as the body (being 31 mm. long when the body is 7 mm.). The second joint is considerably swollen, as is the fourth, which contains the ovary; the fifth joint is longer and thinner than the fourth, and the sixth than the fifth. The first tarsal joint is longer than the second, the auxiliary claws are two-thirds of the length of the claw. The fourth and fifth joints are almost smooth, with the exception of some strong hairs on the fifth joint, and a row of smaller spines at the extremity. On the sixth joint the number of hairs increases towards the extremity; the two last joints have a large number of rather strong hairs.

The leg of a male of 6 mm. is about 25 mm. The leg is a great deal more slender, especially the second and fourth joints, but the relative length of the joints is the same. The auxiliary claws of the males are a great deal shorter than those of the females, their length never reaching half the length of the claw, and often being much shorter. The genital openings of the females are much larger than those of the males. In the females they are found on all the legs, in the males they are wanting on the first pair of legs.

Habitat.—With the foregoing species, south of Halifax.

Station 49. May 20, 1873. Lat. 43° 3′ N., long. 63° 39′ W. Depth, 83 fathoms. Bottom temperature, 1·8° C. Sea bottom, gravel.

Observations.—Of this species eight specimens were dredged by the Challenger. Of these five are females. Of the three males one is furnished with eggs. These specimens agree perfectly as to the relative lengths of the joints of the palpi, and of the tarsal joints of the legs, so that these characteristics are, for this species at least, really distinguishing features ; these marks, together with the structure of the first segment of the body, of the oculiferous tubercle, &c., make this species one of the most sharply-characterised forms of the genus.

Nymphon brachyrhynchus, n. sp. (Pl. IV. figs. 2–7).

Diagnosis.—Body not very slender, smooth ; legs hairy. Proboscis short. Mandibles large. Second joint of palpi a little shorter than third. Second tarsal joint of the leg longer than first. Auxiliary claws small.

Description.—The body of this species is not very slender ; yet there are distinct (but small) intervals between the lateral processes of the body. The proboscis is short, the segment formed by the conjunction of the cephalic and the first thoracic segment is also short, as well as the following thoracic segments. The abdomen is comparatively large and robust. The oculiferous tubercle is small and blunt, the eyes are four in number, small, and not very distinct, light brown. The body is quite smooth. The length of the female is nearly 7 mm., that of the male 6·5 mm.

The mandibles are large. The first joint is almost as long as the proboscis, the second joint curved and long, the claws long also. These claws are curved and furnished with numerous teeth ; they are more numerous and smaller on the immovable than on the movable claw (Pl. IV. fig. 4). The two claws when closed meet along their whole length, the tips only crossing for a small extent.

The palpi (Pl. IV. fig. 5) are very slender, more than twice as long as the proboscis. The third joint is longer than the second, the fourth again longer than the third, the fifth much more slender, but almost as long as the second. Hairs are scarce on the second, not very numerous on the third joint, more numerous on the fourth, and very numerous on the fifth joint.

The ovigerous legs of the males are more than 10 mm. long ; the fourth joint slightly, the fifth strongly curved and very long, the sixth joint short ; the four last joints are short and wound up spirally. Small hairs are placed vertically on the fifth and sixth joints, larger ones at the extremity of the joints. The denticulated spines greatly resemble those of *Nymphon longicoxa*, though they are a little flatter ; their numbers are respectively 13, 9, 7, 7. The spines of the claw are small and not very numerous (Pl. IV. fig. 6).

The ovigerous leg of the female is a great deal smaller : a female of 7 mm. in length, has ovigerous legs of 8·5 mm. The fourth and the fifth joints especially are much shorter, and the latter are quite straight. The denticulated spines of the four last joints are more numerous than in the males ; they are 14, 12, 10, 9, which numbers, however, vary slightly for the different specimens. The claw with its spines is like that of the male.

The legs are slender, those of a male of 6·5 mm. nearly 25 mm., those of a female of 7 mm. nearly as long (25·5 mm.). The second joint is more than thrice as long as the third, the fourth and the fifth nearly of the same length, the sixth almost once and a half the length of the fifth. The two tarsal joints, of which the second is longest, describe a slight curve; the claw is large, the auxiliary claws are very small (Pl. IV. fig. 7). Beginning at the fourth joint the number of hairs regularly increases down to the end of the leg.

The legs of the females show the second and fourth joints considerably swollen. In the females the genital pores are large, and to be found on every leg; in the males they are much smaller, and occur only on the two hind pairs of legs.

Habitat.—A large number of specimens of this species was gathered in Christmas Harbour, Kerguelen. There are a great number of females with the ovaries swollen in the fourth joint of the leg; a great many males with and without eggs or larvæ on their ovigerous legs.

The specimens were dredged at a depth of 45 to 120 fathoms. (January 29, 1874. Off Christmas Harbour. Depth, 120, 105, and 45 fathoms.)

Observations.—This fine species shows very characteristic marks, and cannot easily be confounded with other species. In some respects it shows a resemblance to *Nymphon strœmii* of Kröyer.

Nymphon fuscum, n. sp. (Pl. IV. figs. 8–11).

Diagnosis.—Body and legs very slender. Cephalic segment of the body large, occasioned by the length of the so-called neck. Second joint of the palpi the longest; second joint of the leg three times as long as first; tarsal joints of leg nearly equal. Accessory claws present.

Description.—In this species the body is again extremely slender. The proboscis is short, the mouth small. Of the cephalothoracic segment the so-called neck is long, the segment itself is not very long, there being no interval between the ovigerous leg and the first true leg. Between the true legs the intervals are, on the contrary, very great. The abdomen is minute. The eyes are very large, covering almost the entire surface of the short and blunt oculiferous tubercle.

Of the mandibles, the first joint is longer than the proboscis, the second is comparatively short, and so are the claws. The movable claw is a great deal more slender than the immovable one, the latter is furnished with larger teeth, which are not so acute as the smaller ones of the movable claw (Pl. IV. fig. 9).

Of the palpi the second joint is the largest, the fourth the shortest (Pl. IV. fig. 10). The last joints are extremely hairy, but hairs are also to be found on the second and third joints. The palpi are not very long, but stout; their length being about one-half the length of the proboscis.

The ovigerous legs of the males are very characteristic, forming a very elegant curve. The fifth joint is the longest and the most strongly bent, the sixth joint is a great deal shorter, the four last joints are very short; the claw is short and furnished with very rudimentary teeth (Pl. IV. fig. 11). The spines of the four last joints are very slender, but they are almost all broken at the extremity, their numbers are respectively 14, 14, 13, 13. On the fourth joint hairs are scarce, on the fifth there are a great many small hairs vertically implanted on the surface of the joint, the hairs of the sixth and the four last joints are also numerous : they are larger than those of the fifth joint.

The legs are very slender : in a male of 8 mm. their length is nearly 32 mm. The second joint is more than twice as long as the third, the fifth joint is only a little longer than the fourth ; the sixth is once and a half as long as the fifth. The two tarsal joints are nearly equal. The claw is short, being one-third the length of the last joint of the leg. The accessory claw is not half as large as the claw. The first joints of the leg are nearly smooth, but from the sixth to the eighth they are covered with very small rough hairs.

The colour of this species is a dark brown, much darker than is the case with the other species. There are in all three specimens, of these two are mature males (one furnished with eggs), the third is a very small one with broken accessory legs. I consider it a young male. Genital pores I observed only on the second joint of the two last legs.

Habitat.—This species was found off Kerguelen Island at a depth of 25 fathoms.

· Station 149. January 17, 1874. Lat. 49° 40′ S., long. 70° 28′ E. Off Royal Sound. Depth, 25 fathoms.

Observations.—This species shows a certain resemblance to *Nymphon gracilipes*, Miers. Yet there are too many small differences for me to feel justified in considering the two forms as identical. The form of the oculiferous tubercle is not as figured by Böhm, the claws of the mandibles are shorter and not so straight as in *Nymphon gracilipes*, Miers (after Böhm, Pycnogon. des Museums zu Berlin, Monatsb. der K. Akad. der Wiss. zu Berlin, p. 170, 1879) ; the length of the fourth joint of the accessory legs is different,[1] the length of the two tarsal joints, of the claw, the colour of the whole animal, so characteristic in my *Nymphon fuscum*, is quite different from that described by Böhm (fast weiss bis hellbräunlich, Böhm, *loc. cit.*, p. 172). There can be little doubt, however, that these two forms are closely allied.

Nymphon brevicaudatum, Miers (Pl. IV. figs. 12, 13 , Pl. V. fig. 1–5. *Nymphon hispidum*, n. sp., is marked on the plates).

Nymphon brevicaudatum, Miers, Ann. and Mag. of Nat. Hist., vol. xvi. p. 107, 1875 ; Crustacea of Kerguelen Island, Phil. Trans. Lond., vol. clxviii. (extra vol.) pp. 200–214, pl. xi. fig. 8, 1879.
Nymphon horridum, Böhm, Pycnogoniden des Museums zu Berlin, Monatsber. der K. Akad. der Wiss. zu Berlin, 1879, p. 172, taf. i. fig. 3–3f.

Description.—The body of this species is very robust, and has the lateral processes scarcely separated. The proboscis is comparatively short and bears a small mouth ; the

---

[1] Böhm, *loc. cit.*, taf. i. fig. 1d, represents the sixth joint of the ovigerous leg as furnished with denticulated spines. Of course this is a mistake.

first segment of the body is considerably swollen at the base of the mandibles, constricted posteriorly, and shows a faint line at the beginning of the first true thoracic segment. The abdomen is comparatively long. The upper surface of the body is armed with numerous spines, forming in the middle of the segments star-like groups (Pl. IV. fig. 12), and scattered more irregularly on the lateral processes; near the end of these a row of stronger spines is observed. The oculiferous tubercle is, especially in the females, highly elevated, and shows four dark eyes separated by a cross-like spot. The oculiferous tubercle is placed on the cephalic part of the first segment, between the two ovigerous legs.

The mandibles are very long, the first joint being much longer than the próboscis; the second joint is shorter and strongly curved, the claws being placed almost transversely before the mouth. The immovable claw is more strongly curved than the movable one; the spines on the claws are numerous, but they are almost equal and of the same length. Both joints are covered with numerous small hairs, stronger ones being found at the end of the first joint (Pl. V. fig. 2).

The palpi are not very hairy. The second joint is by far the longest, the fourth is the shortest, the fifth very slender. The whole length of the palpus is about twice the length of the proboscis (Pl. V. fig. 1).

The ovigerous legs of the males are not quite so long as the body. The fifth joint is the longest, and is considerably swollen at the extremity, the sixth joint is a great deal shorter, the seventh to the tenth joints are very short, the claw almost of the same length as the last joint. The ovigerous legs are very strongly bent and not very hairy. The denticulated spines of the four last joints are sharply serrated and not numerous, their numbers being respectively 3, 3, 2, 3 (Pl. V. figs. 3, 4). The claw is furnished with a row of acute spines.

The ovigerous legs of the females are still shorter than those of the males. The relative length of the joints is the same, but the fifth joint is not swollen at the extremity. The leg is not curved as is the case with that of the male.

For a species of *Nymphon* the true legs are uncommonly short: a female of 6 mm. has a leg of only 14 mm. The second joint is not quite twice as long as the first or third, the three following joints are almost of the same length and comparatively robust, especially the fourth joint of the female. The two tarsal joints, the second of which is the longer, are very slender and almost smooth, together about as long as the sixth joint. The claw is about as long as one-third of the length of the second tarsal joint, the auxiliary claws are very small. The legs are very hairy. These hairs vary greatly in size and strength, and are not placed in regular rows. The dorsal surface of the leg especially is covered by a large quantity of stronger spines. The genital pores of the females are very large and visible on the second joint of every leg. Those of the males are small, and found only on the second joints of the two last legs.

I wish to point out as a very interesting peculiarity of this species, that I observed a female specimen with highly developed ovaries in the fourth joint of its legs, which bears a distinct egg-mass on its ovigerous leg. The shape of the ovigerous leg is almost identical with that of a male. In other respects it looks quite like a female.

Habitat.—This species seems to abound in the neighbourhood of Kerguelen Island. Miers (*loc. cit.*, p. 213) only says that several specimens were collected at this island, but Böhm (*loc. cit.*, p. 177) mentions specimens collected in Royal Sound, Irish Bay, and Great Whale Bay, from a depth of 15 to 65 fathoms.

The Challenger brought specimens home from Station 149. January 17, 1874. Lat. 47° 40′ S., long. 70° 20′ E. Off Royal Sound. Depth, 25 fathoms.

January 20, 1874. Royal Sound. Depth, 28 fathoms.

January 29, 1874. Off Christmas Harbour. Depth, 120 fathoms.

Observations.—After a close examination of the numerous specimens of this Pycnogonid, I at first believed it to be different from the *Nymphon horridum* of Böhm, so I gave it the name *Nymphon hispidum*, which name is still to be found on Plates IV. and V. of this Report. After a renewed examination, and having acquired, I believe, by continuous study some knowledge of the relative value of characteristic marks, I think there can be no doubt about the identity of Böhm's species and the specimens brought home by the Challenger. Unfortunately the two plates are printed off, and thus bear the name originally proposed by me.

However, the name proposed by Böhm ought also to give place to another, viz., that of Miers. The description of Mr Miers originally published was too short and insufficient, therefore Mr Böhm was quite justified in considering his specimens as distinct and proposing for them the new name *Nymphon horridum*. This happened in 1879. In the same year Mr Miers published a more detailed description with figures, which made it certain that Böhm's, Miers', and my specimens belonged to the same species; this must, I think, bear the name originally applied to it by Mr Miers.

The latest description of this author, however, is by no means exhaustive; his figures are very small, and when he says that the number of claws at the end of the leg is two, it is evident that he has not studied the details with a high enough power.

Böhm's description and drawings are much superior to those of Miers; they differ from mine in the following respects:—On the dorsal surface of the body he figures rows of hairs between the different segments, whereas I observed star-like groups of hairs. He gives the mandibles a much more elongated form, and furnishes the claws with very irregular teeth. On the legs, Böhm says, the hairs and spines are arranged in regular rows, whereas I failed to observe this regularity. The two tarsal joints as figured by Böhm are nearly of the same length, but I always observed that the second was much longer than the first. Minor differences in the form of the eyes, distribution of the hairs, &c., it is unnecessary to discuss. The more important ones which I have pointed out must perhaps partly be attributed to inaccuracy ; they may have been occasioned by the circumstance that Böhm's material was much more limited than mine. Böhm doubts whether he has a male example or not, whereas I had a dozen at my disposal, five of which were furnished with eggs.

Nymphon perlucidum, n. sp. (Pl. V. figs. 6–10).

Diagnosis.—Body and legs very slender, pellucid and smooth. Second joint of the palpi elongated, much longer than the third. Second joint of the feet more than twice as long as the third. First tarsal joint uncommonly short. Accessory claws.

Description.—This is a very small and a very fine species, the most transparent form of *Nymphon* I ever observed. The proboscis is robust, yet very long, much longer than the first segment of the body. There is no interval between the lateral process of the ovigerous leg and that of the first true leg, but the intervals between the other lateral processes are very large. The abdomen is small. The eyes are obsolete : a very small tubercle without pigment is all that is to be seen.

The mandibles have the first joint as long as the proboscis, the second small with very long claws, the immovable claw is strongly curved at the extremity. The number of spines on these claws is much more limited than in any of the other species of *Nymphon*, being four on the movable claw and only five on the immovable one. The mandibles are scarcely hairy : a few stronger hairs are observed at the end of the first joint, and again at the origin of the movable claw.

The palpi are not very long, once and a third as long as the proboscis. They are slender, the second joint being comparatively much longer than the third joint, and nearly as long as the three last joints together. They are covered with very small hairs.

The ovigerous legs are short : 4½ mm. in a male of 5 mm. ; the fifth joint is the longest, considerably swollen at the extremity ; the sixth is a great deal shorter and feebly bent ; the four last joints are very short, the claw being half the length of the last joint. The different joints are covered all over with very minute hairs. The spines of the four last joints are broad, but very small and almost show the hand form (Pl. V. fig. 9). They are not very numerous, their numbers being respectively 7, 4, 4, 5. The claw is furnished with some small spines (Pl. V. fig. 8).

The legs are slender, but not very long, only two and a half times as long as the body. The second joint is twice as long as the first and third ; the fifth is longer than the fourth, and the sixth is the longest. The two tarsal joints are highly characteristic on account of the shortness of the first, and the strong spines placed along that side of the second which is opposite the claw. ·The length of the accessory claws is not half the length of the claw. The legs are covered with small hairs, stouter ones being placed on the fifth and sixth joints, and at the end of the joints. I have figured the two tarsal joints in fig. 10, Plate V.

The only specimen of this species collected by the Challenger is a male furnished with genital pores on the second joint of the two last legs.

Habitat.—This small species was dredged during the Challenger Expedition between Celebes and Halmahera.

Station 196. October 13, 1874. Lat. 0° 48′ S., long. 126° 58′ E. Depth, 825 fathoms. Bottom temperature, 2·4° C. Sea bottom, rock.

Observations.—This fine species is highly interesting, being among the slender species of *Nymphon*, the only one in which the form of the two last joints of the leg shows a remarkable likeness to that of the same joints of most other genera of Pycnogonida. This, however, is not the only characteristic point; a second is that the claws of the mandibles are not armed with a row of very numerous spines as in the other species of *Nymphon*, but only with three, four, or five spines. The species is blind, yet the depth from which it was brought up was only 825 fathoms, whereas *Nymphon meridionale*, Hoek, *e.g.*, lives at a depth of 1674 fathoms and has four distinct eyes.

Ascorhynchus, G. O. Sars.

Ascorhynchus glaber, n. sp. (Pl. VI. figs. 5–9; Pl. XV. fig. 16).

Diagnosis.—Proboscis one-third of the total length of the body. Abdomen half as long as the proboscis. Body and legs almost entirely smooth, with the exception of three strong spines placed dorsally on the hinder margin of the first three thoracic segments.

Description.—

Length of the proboscis, ...	8½ mm.
Length of the trunk, ...	13½ ,,
Length of the abdomen, . :	4 ,,
Total length of the body, ...	26 ,,
Length of the cephalothoracic segment, ...	7 . ,,
Length of the third leg, ...	39 ,,

The body of this beautiful species is strong, yet comparatively slender, with great intervals between the lateral processes. The proboscis is very stout, pyriform, distinctly triangular in transverse section; each of the three sides of the proboscis is longitudinally furrowed in the middle; at its extremity the proboscis is sharply pointed, the mouth is small and triangular. The proboscis is distinctly divided into a fore, middle, and hinder part, the latter tapering towards the extremity, where it articulates with the cephalothorax (Pl. XV. fig. 16). The length of the cephalothoracic segment is very considerable, being about half that of the trunk. Anterior to the insertion of the palpi it is a little narrower, and at the front part it bears the mandibles, between which it shows a small azygous knob. On the dorsal surface a slight elevation is observed between the two palpi, whereas behind the middle, almost exactly between the two ovigerous legs, the same surface bears a very high conical oculiferous tubercle, which as a sense organ is quite rudimentary, being destitute of lenses, pigment, &c. A second dorsal conical elevation is found at the hinder margin of the cephalothorax, and similar very strong spines are also observed at the hinder margin of the two following thoracic segments (Pl. VI. fig. 5). At the point where these thorns arise the segments are a great deal wider than the anterior part of the following segment. The abdomen is comparatively long but very narrow, being only a little swollen at the extremity. The lateral processes for the attachment of the ovigerous legs are short, those for the true legs comparatively very long.

Both specimens of this species brought up by the Challenger are furnished with three-jointed mandibles. Those of the younger specimen bear at the extremity of the third joint slender and curved claws (Pl. VI. fig. 7); the older specimen, on the contrary, shows rudimentary straight and very small claws, the movable claw being furnished with a slender thread (Pl. VI. fig. 6). The latter specimen is a male, and there is no reason why it should not be considered as full-grown.

The form of the palpi is nearly the same as in the other species of *Ascorhynchus*. The first two joints are extremely small, the third is the longest, the fourth is short, the fifth about twice as long, and narrow in the middle, the sixth is very short; of the seventh to the tenth joints, the first is the shortest, the second the longest. With the exception of some extremely small hairs on the last joints, the surface of the palpi is quite smooth.

The ovigerous legs have the fourth joint the longest, the fifth shorter and swollen towards the extremity, the sixth about half as long as the fifth and yet more swollen towards the extremity; of the four last joints the first is the longest, and the claw at the end of the tenth joint is extremely small. The denticulated spines are placed in different rows, each row showing spines of about the same size, whereas those of different rows vary greatly (Pl. VI. fig. 8).

Of the legs the first pair is a great deal less developed than the three following; its total length is only 30 mm., whereas that of the third pair is fully 39 mm. Of the latter leg the second joint is twice as long as the first or third joint, the fourth joint is the longest, the fifth joint is a little shorter than the fourth, the sixth again a little shorter than the fifth; of the two tarsal joints the first is a little shorter than the second, the claw is not quite half as long as the second tarsal joint. There are no accessory claws (Pl. VI. fig. 9). The claw of the first leg is extremely minute, yet distinct. The legs are almost hairless; yet the fourth and the fifth joints cannot be called smooth, as they are furnished with rows of knobs, corresponding with the knobs I observed on the leg of *Nymphon hamatum*, Hoek.

The animal from which all the above measurements, &c., are taken is a male; its genital orifices are found ventrally on the second joint of the two posterior legs. Its colour is a beautiful orange-yellow.

Habitat.—This interesting species was dredged during the Challenger Expedition at Station 146. December 29, 1873. Lat. 46° 46′ S., long. 45° 31′ E. Depth of the sea, 1375 fathoms. Bottom temperature, 1·5° C. Sea bottom, globigerina ooze.

Observations.—This species is, I believe, closely allied to *Ascorhynchus abyssi*, G. O. Sars. It can, however, easily be distinguished from that species.

1. By the proboscis, which is as long as the trunk in *Ascorhynchus abyssi*, and only two-thirds the length of the trunk in *Ascorhynchus glaber*.

2. By the lateral processes of the body, which in *Ascorhynchus abyssi*, " vix," in *Ascorhynchus glaber* are visibly longer than the breadth of the body.

3. By the oculiferous tubercle, which in *Ascorhynchus abyssi* is denticulate and placed on the fore part of the first segment, while in *Ascorhynchus glaber* it is behind the middle of the cephalothorax and quite smooth.

4. By its three-jointed mandibles.

5. By the legs, which are covered with short hairs and·are twice as long as the body in *Ascorhynchus abyssi*; in *Ascorhynchus glaber*, on the contrary, they are almost smooth, and are only once and a half as long as the body.

6. By the length of the body, 10 mm. in *Ascorhynchus abyssi*, and 26 mm. in *Ascorhynchus glaber.*

Ascorhynchus minutus, n. sp. (Pl. VI. figs. 10–16).

Diagnosis.—Proboscis not quite one-third of the total length of the body. Abdomen one-third the length of the proboscis. Body and first joints of the legs furnished dorsally with numerous strong spines.

Description.—

Length of the proboscis,	.	.	.	.	2	mm.
Length of the trunk,	.	.	.	.	.	3·75 ,,
Length of the abdomen,	.	.	.	.	.	0·65 ,,
Total length of the body,	.	.	.	.	.	6·4 ,,
Length of the cephalothoracic segment,	.	.	.	2·3 ,,
Length of the third leg,	.	.	.	.	.	9·0 ,,

The body of this nice little species is slender, and the intervals between the lateral processes are great (Pl. VI. fig. 10). The body and legs are almost entirely smooth, single hairs being found only at the distal extremities of the joints of the legs. The proboscis is triangular, almost of the same form as that of *Ascorhynchus glaber;* it is distinctly divided into three parts, the middle part being considerably swollen. The cephalothoracic segment is comparatively long, being as long as the remaining part of the body including the abdomen. The oculiferous tubercle is situated in front of the middle of the cephalothoracic segment and is much elevated; the eyes are rudimentary. A distinct knob is to be seen at the front margin of the segment between the origin of the two mandibles. The three following thoracic segments are short; the two middle segments are furnished like the cephalothoracic segments dorsally at their hinder margin with an elevated prickle. The lateral processes which serve for the insertion of the legs, are of considerable length; at their extremities they bear a strong prickle, which is not quite so long as those on the middle of the dorsal surface.

The mandibles consist of a single joint, bearing at its extremity a rudimentary triangular second joint.

The palpi are slender and comparatively long; the first two joints are extremely small, the third joint is the longest; the relative length of the other joints is nearly the same as in the palpi of *Ascorhynchus glaber*. The distal extremity of the fifth joint and the whole surface of the four following joints show on the one side numerous hairs of the curious form observed and described by Böhm for *Ascorhynchus ramipes*, Böhm (sp.) (*Gnamptorhynchus ramipes*, Böhm) (Pl. VI. fig. 16).

The ovigerous legs are comparatively short, 5 mm. in the specimens of 6·4 mm. The first joint is very small, the two following are a little longer, the fourth and fifth are very long, the sixth a great deal shorter; the four last joints (Pl. VI. fig. 12) are very short, and about the same length, the claw is not so extremely short as in *Ascorhynchus glaber*. The denticulated spines are placed in three distinct rows (Pl. VI. fig. 13).

The first joints of the legs are furnished dorsally with strong prickles, like those of the dorsal surface of the body and the lateral processes. The first joint of the leg is furnished with two, the second joint, which is but little longer than the first, with a single prickle. On the third joint, which is nearly as long as the first, no prickles are to be seen. The fourth is shorter than the fifth; the latter, the longest of all, is also a little longer than the sixth joint. Of the two last joints the first is a little shorter than the second (Pl. VI. fig. 14). The claw is longer than half the length of the second tarsal joint. Accessory claws are wanting. The first pair of legs, which is much feebler than the three following pairs, has a very small claw (Pl. VI. fig. 15).

Both specimens of this species collected during the cruise of H.M.S. Challenger are males. I observed small genital pores on the two hinder legs at the place which, as far as my knowledge goes, they invariably occupy. One of these males bears larvæ, for the description of which see below.

Habitat.—The specimens of *Ascorhynchus minutus* were dredged in the neighbourhood of Melbourne, at

Station 161. April 1, 1874. Off entrance to Port Philip. Depth, 38 fathoms. Sea bottom, sand.

Observations.—This species seems to be nearly allied to *Ascorhynchus ramipes*, Böhm (sp.). Yet it may be easily distinguished from it by the following characteristics :—

1. The total length of the body of *Ascorhynchus ramipes* is 11 mm., of the proboscis

3 mm., whereas the length of the body of *Ascorhynchus minutus* is only 6·4 mm., and that of the proboscis 2 mm.

2. The abdomen of *Ascorhynchus ramipes* is about as long as the proboscis, while that of *Ascorhynchus minutus* is nearly one-third the length of the proboscis.

3. On the dorsal surface the thoracic segments and the lateral processes in *Ascorhynchus ramipes* show round knobs, in *Ascorhynchus minutus* distinct prickles.

4. The four last joints of the ovigerous legs of *Ascorhynchus ramipes* are furnished with a single row of denticulated spines; those of *Ascorhynchus minutus* with three distinct rows. (I think this difference is not a real one : Böhm's observation and description will, no doubt, in this respect, be found deficient.)

5. The first true leg of *Ascorhynchus ramipes* is not furnished with a claw, that of *Ascorhynchus minutus* has a small, but distinct claw.

Ascorhynchus orthorhynchus, n. sp. (Pl. V. figs. 11–13 ; Pl. VI. figs. 1–4 ; Pl. XV. figs. 14, 15).

Diagnosis.—Proboscis almost half the length of the body. Abdomen not quite one-third the length of the proboscis. Dorsal surface with a row of prickles, also the lateral processes.

Description.—

Length of the proboscis,	.	.	.	.	.	.	10 mm.
Length of the trunk,	.	.	.	.	.	.	9½ ,,
Length of the abdomen,	.	.	.	.	.	.	3 ,,
Total length of the body,	.	.	.	.	.	.	22½ ,,
Length of the palpi,	.	.	.	.	.	.	14 ,,
Length of the third leg,	.	.	.	.	.	.	50 ,,
Length of the ovigerous leg,	.	.	.	.	.	.	20 ,,

The body of this species is very slender and almost smooth. The proboscis does not incline to the ventral side, and is not so much swollen as in the other species of *Ascorhynchus*, consequently the form is not pyriform, but rather club-shaped. The proboscis shows longitudinal furrows ; the mouth is, as in the other species, triangular and very small (Pl. XV. fig. 14).

The cephalothoracic segment is comparatively short, being only half as long as the proboscis ; the mandibles and the palpi are placed on distinct prominences. The oculiferous tubercle is elevated and conical, and furnished with four distinct eyes ; the lateral processes for the ovigerous legs are very small, those for the first pair of true legs as well as those for the following legs are of considerable size. The lateral processes of the true legs are widely separated, thus contributing to the slender appearance of the whole body. Posteriorly the cephalothoracic, as well as the two following segments, are considerably elevated dorsally, while the anterior part of the following segment is much narrower, and placed lower; these three elevations bear in their centre strong conical prickles (Pl. V. fig. 11). Similar prickles, though a little smaller, are also to be observed dorsally on the lateral processes close to the margin of the articulation with the leg.

The mandibles of the female specimen (the only one procured) are small, but distinctly three-jointed. The first two joints are extremely slender and of equal length; the third joint (Pl. VI. fig. 1) is very small, and furnished with rudiments of claws only.

The palpi are very slender, and ten-jointed. The first joint is small, the second—correctly observed by Böhm in *Ascorhynchus ramipes*—much smaller still, the third the longest of all, the fourth small, the fifth not quite half as long as the third, the sixth about the same length as the fourth, and the seventh twice as long as the sixth; the three last joints decrease regularly in length. The first four joints are nearly smooth, distinct hairs are observed towards the extremity of the fifth joint, and on the five following joints (Pl. V. fig. 13). These hairs exhibit the curious form observed by Böhm in the hairs of the palpi of *Ascorhynchus ramipes*; they are also found in *Ascorhynchus minutus* (Pl. VI. fig. 16).

The first three joints of the ovigerous legs are small, the fourth is the longest; the fifth and sixth, which are a great deal smaller, and the seventh, which is extremely small, are quite straight. The last three joints are placed at right angles to the foregoing. All the joints are almost entirely smooth, with the exception of a few hairs towards their extremities. The claw is very small. The denticulated spines on the four last joints are placed in three or four rows; their form can be understood from the drawing in fig. 3 of Plate VI.

The legs are more than twice as long as the body; the second joint is twice as long as the first, the third is as long as the first. The fourth joint is considerably swollen, the fifth nearly as long but a great deal narrower than the fourth, the sixth much narrower and a little longer also. Of the two tarsal joints the second is longer than the first and considerably curved, and the claw is nearly half as long as the second tarsal joint. The hairs on the legs are very small, and can only be seen with the microscope. Some stronger hairs are placed at the distal ends of the joints.

The single specimen brought up by the dredge is a female (having the fourth joint of the legs swollen and the ovigerous legs feeble). The genital pores are not very small, and are found ventrally at the end of the second joint of every leg (Pl. XV. fig. 15). Judging from the whole exterior of the animal, it is a full-grown, or nearly full-grown, specimen.

Habitat.—The only specimen of this species was dredged north of New Guinea, at Station 219. March 10, 1875. Lat. 1° 50′ S., long. 146° 42′ E. Depth of the sea, 150 fathoms. Sea bottom, mud.

Observations.—This curious form of *Ascorhynchus* does not show any remarkable affinity with any of the other forms of *Ascorhynchus* described. Judging from the shape of the proboscis, it comes nearest to some of the species of *Colossendeis*. Yet in that genus the mandibles in the full-grown animal have totally disappeared, whereas *Ascorhynchus orthorhynchus* in all probability has these appendages in the adult state. I do not believe, however, as I have said before, that this difference is in reality very important, especially since I have observed among the specimens of *Colossendeis gracilis* one furnished with long three-jointed mandibles, while these appendages were totally wanting in the other specimens of the same species.

Oörhynchus, n. gen.

Diagnosis.—Proboscis ovate, inserted ventrally on the cephalothorax at a considerable distance from the front margin. Mandibles rudimentary; palpi nine-jointed. Ovigerous legs ten-jointed, the four last joints not furnished with one or more rows of denticulate spines.

Oörhynchus aucklandiæ, n. sp. (Pl. VII. figs. 1–7).

Diagnosis.—Oculiferous tubercle horizontally directed forwards. Mandibles represented by single-jointed club-shaped bodies. First tarsal joint of the legs extremely small, auxiliary claws wanting. Abdomen very long, once and a half as long as the proboscis.

Description.—

Length of the proboscis,	. . . . . .	1 mm.
Length of the cephalothorax,	. . . . .	0·9 „
Length of the trunk,	. . . . . .	2 „
Length of the abdomen,	. . . . . .	1·5 „
Total length of the body,	. . . . . .	4·2 „
Length of the leg,	. . . .	5·5 „

This very curiously-shaped Pycnogonid has the cephalothoracic segment short but very broad, furnished at the front with a long cylindrical oculiferous tubercle which projects horizontally beyond the extremity of the proboscis. The oculiferous tubercle is furnished with four eyes, two placed dorsally, and two ventrally; the latter two are the smaller. The cephalothorax is armed at the two corners with curiously-shaped spines also projecting forwards, and above the attachment of the first pair of legs bears a couple of long hairs placed on small knobs. Similar pairs of hairs or thin spines are also observed on the two following thoracic segments on the dorsal surface between the lateral processes for the insertion of the legs. The rest of the surface of the body is entirely smooth. The three thoracic segments are small, and the lateral processes are separated by small intervals. The abdomen, on the contrary, is very long, being once and a half as long as the proboscis. The abdomen shows on both sides a row of comparatively long and projecting hairs.

The proboscis is inserted ventrally on the cephalothorax, at a considerable distance from the front margin ; it is of a distinctly ovate form, and bears the small triangular mouth surrounded by slightly swollen lips.

The mandibles are represented by one-jointed robust club-shaped bodies, placed at the front margin of the cephalothorax on both sides of the oculiferous tubercle, but not reaching quite so far as that organ. The mandibles are covered all round with strong hairs, placed at right angles to the surface, and when the mandibles are viewed from below and anteriorly, the round cicatrice is observed where in all probability, at an earlier date, a second joint has been inserted.

The palpi are nine-jointed and placed close to and on both sides of the base of the proboscis. The first joint is very short, the second is the longest of all and directed forwards, the third again is short, the fourth almost as long as the second, and bent so as to form an angle with the first three joints. The fifth joint is again very short, the four last joints are nearly of the same length ; the sixth, however, is the longest, and the eighth the shortest. All the joints are furnished with very strong hairs, which are longest on the third and fourth joints, and decrease in size though not in number towards the extremity of the appendages.

The ovigerous legs are strongly curved. Of the first three joints, which are not so very small as is the case in other genera, the second is the longest, being nearly twice as long as the first. The fourth and fifth joints are nearly of the same length and are the longest of all. The sixth is not quite half as long as either of the two foregoing joints ; of the seventh to the tenth the first is the longest, the second much shorter and the third a little shorter, while the last joint is extremely small and may easily be overlooked. The first three joints are almost smooth, the two following are furnished with very small hairs, and the sixth to the tenth with not very numerous but longer hairs. The last joint but one is armed with one, the last with two not very strong, comparatively broad, and slightly serrated spines. The total length of the ovigerous leg is not quite so long as that of the body without the abdomen.

The legs are comparatively strong and very hairy (Pl. VII. fig. 6). The lateral processes are furnished with numerous small and curved spines ; the first three joints of the leg are small and nearly of the same length. The fourth joint is twice as long as the third, and on the dorsal surface beyond the middle it bears a tubular process, directed backwards towards the origin of the leg. The fifth and sixth joints are the longest in the leg, and are nearly of the same length. The fifth joint, however, is a great deal more slender than the fourth, and the sixth much more so than the fifth. The first tarsal joint is extremely small, the second long and distinctly curved ; the claw is scythe-shaped, and accessory claws are wanting. All the joints are furnished with long and strong hairs, standing at right angles to the surface ; the last joints are covered with much more delicate hairs.

The only specimen of this curious species dredged by the Challenger Expedition is a male; the second joints of the two last legs bear on the ventral side distinct rounded tubercles, and at the tips of these knobs the small genital pores are to be observed.

Habitat.—*Oorhynchus aucklandiæ* was dredged at Station 169. July 10, 1874. Lat. 37° 34′ S., long. 179° 22′ E. Depth, 700 fathoms. Temperature of the bottom, 4·2° C. Sea bottom, grey ooze.

No other species of Pycnogonid was dredged at this station.

Observations.—No doubt the genus *Oorhynchus* is nearly allied to other genera of the same group, and especially to the genus *Achelia*. Although the shape of the body is widely different, we find in this genus likewise rudimentary mandibles, ovigerous legs furnished with rudimentary denticulated spines and with 'the four last joints, in general, almost of the same shape as those of *Oorhynchus*. In both genera the genital pores of the males are placed on tubercles situated ventrally on the second joints of the two posterior legs. They are distinguishable by the shape of the body, which is much more concentrated and rounded in *Achelia*, by the number of joints in the palpi, and by the absence of auxiliary claws from the legs of *Oorhynchus*, whereas all the species of *Achelia*, as far as is known at least, are furnished with them.

Colossendeis, Jarzynsky.

Colossendeis gigas, n. sp. (Pl. VIII. figs. 1, 2 ; Pl. X. figs. 1–5).

Diagnosis.—Proboscis bottle-shaped ; cephalic part of the cephalothoracic segment triangular and distinct ; eyes obsolete ; third joint of the palpus longer than the fifth, palpus as long as the body ; claws of the legs minute.

Description.—

	No. 1.	No. 2.	No. 3.	No. 4.
Length of the proboscis,	47·5 mm.	29 mm.	15 mm.	22 mm.
Length of the trunk (with the abdomen),	32·5 ,,	20·5 ,,	14 ,,	13 ,,
Length of the abdomen,	6 ,,	3·5 ,,	2 ,,	2·7 ,,
Length of the palpus,	81 ,,	53 ,,	?	38·5 ,,
Length of the ovigerous leg,	137 ,,	90 ,,	49 ,,	60 ,,
Length of the leg of the third pair,	301 ,,	211 ,,	112 ,,	143 ,,

The body of this gigantic Pycnogonid is robust; nevertheless there are distinct intervals between the lateral processes. In the large specimen (No. 1) the surface of the body is quite smooth; the palpi alone are furnished with strong hairs, and the ovigerous legs with small ones, while the extremely small hairs on the legs can only be seen with a lens. In the younger specimens the hairs are by no means so scarce. Those on the trunk, the proboscis, and the lateral processes are still very small, but the hairs on the legs are much more distinct, and, especially at the distal extremities of the joints of the legs, rows of short strong hairs may be observed ; finally, the palpi and the ovigerous legs show in these younger animals strong but not very long hairs projecting vertically from the surface.

The proboscis is bottle-shaped and very long. The lower stalk-like and more slender part (the neck of the bottle) can easily be distinguished from the middle part which is considerably swollen, whereas the anterior part is a little narrower again, but by no means so narrow as the stalk-like part. The mouth is triangular and very large. The cephalic part of the cephalothorax in this species is sharply divided from the thoracic part, the latter forming a true first thoracic segment. The cephalic part has, when viewed dorsally, a triangular shape; it bears almost exactly in the middle a very blunt oculiferous tubercle. In the younger specimens this tubercle is conical, much more elevated and pointed at the extremity; it shows neither pigment nor lenses.

The four thoracic segments are closely united, so that their terminations are not visible; the origin of the lateral processes for the insertion of the legs is, on the contrary, distinctly marked by a line.

The abdomen is club-shaped and not longer than $\frac{1}{15}$th of the total length of the body. As in the other species of the genus *Colossendeis* it is connected with the thorax by means of an articulation.

The palpi are as long as, or a little longer than, the body. The two first joints are extremely small, the third is very long, the longest of all the joints of the palpi; it is a little swollen at its beginning and again at the extremity. The fourth joint is again very small, the fifth almost as long as the third; the sixth is again much shorter, nearly one-fourth the length of the fifth. Of the last four joints the seventh is the longest, the eighth being only half its length, and the ninth and tenth, which are of equal length, being still shorter than the eighth. The first four joints are almost quite smooth, while the following joints are furnished with short but rather strong hairs.

The ovigerous legs, as in the other species of this genus, are attached close to the base of the palpi. The lateral processes for the ovigerous legs are found at the ventral side, immediately behind and close to the first joint of the palpi. The first three joints of the ovigerous legs are extremely short, not longer than they are broad. The fourth joint is very long (almost as long as the proboscis) and not inconsiderably swollen at the extremity. The fifth joint is again short, and this joint serves—as was observed by Wood-Mason [1]—to elbow the ovigerous leg. The sixth is still longer than the fourth and exactly as long as the proboscis. The last four joints gradually become more slender; they are nearly equal in length and very short, the length, however, decreases a little from the seventh to the tenth joint. At its extremity the last joint bears a very small hook-like curved claw, the inner surface of which is quite smooth.

[1] Wood-Mason, On *Rhopalorhynchus kröyeri*, Journ. of the Asiatic Soc. of Bengal, Calcutta, vol. ii., 1873.

The last four joints describe a spiral curve; so that the tenth lies parallel to the seventh. Their inner surfaces are furnished with numerous rows of very flat spines. In the oldest specimen there are about twelve rows, making the total number of spines for every joint several hundreds. None of these spines, however, show the original shape; they are all of them broken by use or by age, and those placed towards one side especially are very short and rudimentary. On the other side they grow longer (Pl. X. fig. 2), and, covered by this outermost row of longest spines, some short thimble-shaped knobs (fig. 5) are to be observed. Such is at least the condition of the flattened spines on the last four joints of the ovigerous legs of the gigantic male dredged at Station 146. The specimen second in size is a great deal smaller and in all probability is not quite adult. Here the spines, placed in about eight rows, show a much more regular shape (fig. 3); their margins are furnished anteriorly with very small hairs. These hairs are rather firm, are not at all injured by the action of alcohol, and must not, I believe, be considered as cilia. The spines are in the middle a great deal narrower, and broader again at the extremity, the broader part has the shape of a rhomb. In the earliest stage the spines are much smaller and beautifully spatulate. The small hairs extend here to beyond the middle (Pl. X. fig. 4).

This species has exceedingly long legs. The first three joints are very small, the three following very long; the fourth is the longest, the fifth a little shorter, the sixth again a little shorter; the two last joints are small again; the second tarsal is only half as long as the first. The claw measures about one-third the length of the second tarsal joint; there are no auxiliary claws. The joints of the leg, from the fourth to the eighth, gradually decrease in thickness. The hairs, which on the legs of the large adult specimen are extremely small and sparse, are stronger and more numerous on the legs of the younger specimens.

In regard to the sexes of the specimens of this species dredged during the voyage of the Challenger, I am sure only of the gigantic specimen. This is a male; it shows genital openings on the ventral surface not very close to the distal margin of the second joint of the two posterior pairs of legs. About the other specimens, whose genital openings I failed to observe, I am in doubt whether they are young males or females. The colour of the specimens is light yellow, nearly the same as that of all other Pycnogonids preserved in spirits. The large full-grown male, however, is of a much darker orange-red colour, with beautiful red bands over the proboscis, at the extremity of the joints of the legs, &c.

Habitat.—*Colossendeis gigas*, seems to occur in different parts of the southern ocean; it was dredged at

Station 146. December 29, 1873. Lat. 46° 46' S., long. 45° 31' E. Depth of the sea, 1375 fathoms. Temperature of the bottom, 1·5° C. Sea bottom, globigerina ooze.

Station 147. December 30, 1873. Lat. 46° 16' S., long. 48° 27' E. Depth of the sea, 1600 fathoms.   Temperature of the bottom, 0·8° C.   Sea bottom, globigerina ooze.

Station 300.   December 17, 1875.   Lat. 33° 42′ S., long. 78° 13′ W.   Depth 1375 fathoms.   Bottom temperature, 1·5° C.   Sea bottom, globigerina ooze.

Of all the species of this genus known at present, this species of *Colossendeis* shows by far the greatest affinity with *Colossendeis leptorhynchus*, which I shall describe further on.   The latter species, however, is much more slender.

Colossendeis leptorhynchus, n. sp. (Pl. VIII. figs. 3–7).

Diagnosis.—Proboscis almost cylindrical.   Cephalic part of the cephalothoracic segment distinctly separated from the thoracic part; eyes obsolete; third joint of the palpus shorter than the fifth; palpus much longer than the body.   Legs and proboscis extremely slender.   Claws of the legs minute.

Description.—

	Male.	Female.
Length of the proboscis, .	28 mm.	33 mm.
Length of the trunk with the abdomen, .	13 „	14 „
Length of the abdomen, .	2·2 „	2·5 „
Length of the palpus, .	35 „	37 · „
Length of the ovigerous legs, .	56 „	57 „
Length of the leg of the third pair, .	143 „	158 „

The body, and especially the proboscis and legs, are a great deal more slender than in *Colossendeis gigas*.   The intervals between the lateral processes are not quite so wide as the thickness of these processes.   The surface of the body is quite smooth, but the palpi and ovigerous legs, and the last five joints of the legs, are furnished with extremely small hairs.

The proboscis is extremely long, more than twice as long as the trunk; about the middle it is slightly swollen.   The cephalic part of the cephalothorax is distinctly separated from the thoracic part.   It is elongated, a little more slender towards the end, and bears on the dorsal surface, about the middle, a small blunt knob as a rudimentary oculiferous tubercle.

The four thoracic segments are closely united.   The abdomen is small, about 1-18th of the total length of the body.   The palpi are not very long.   The first two and the fourth joints are extremely small; the third is long, and the fifth a great deal longer still.   The sixth joint is shorter than the seventh, and of the last three joints (fig. 5), the third is by far the longest.   The relative lengths of the joints of the ovigerous legs is the same as in *Colossendeis gigas*, Hoek; the sixth joint is again considerably longer than the fourth, and the fifth is extremely short.   The last four joints decrease in length, and the claw is very small (fig. 6).

The first three joints of the legs are very small, nearly as long as broad; the following three are very long and slender, and not inconsiderably curved. Their relative length is the same as in *Colossendeis gigas*. Of the last two joints, the first is longer than the second. The claw is small.

Of this species seven adult and two young specimens were dredged during the cruise of H.M.S. Challenger. By stretching out the legs parallel with each other, the body of the animal assumes a very peculiar aspect.[1] Of the eight adult specimens, seven are females—one only a male. Both the females and the male have genital openings only on the two hindermost legs, where they are placed on the ventral surface of the second joint of the leg. Perhaps the ovigerous legs of this species, and in that case probably of the other species of this genus also, have lost their egg-bearing function. Small capsules filled with numerous eggs were attached to several joints of the legs, but as these eggs were quite undeveloped, it was impossible to determine whether they belong really to this animal, or to some other inhabitant of the same locality. The capsules, I observed, were attached to the fourth joint of the leg in a female, and to the second joint of the leg in the single male specimen.

Habitat.—This species was dredged at the same stations as the foregoing species (*Colossendeis gigas*), and also at a fourth station (310).

Station 146. December 29, 1873. Lat. 46° 46′ S., long. 45° 31′ E. Depth of the sea, 1375 fathoms. Temperature of the bottom, 1·5° C. Sea bottom, globigerina ooze.

Station 147. December 30, 1873. Lat. 46° 16′ S., long. 48° 27′ E. Depth of the sea, 1600 fathoms. Temperature of the bottom, 0·8° C. Sea bottom, globigerina ooze.

Station 300. December 17, 1875. Lat. 33° 42′ S., long. 78° 13′ W. Depth of the sea, 1375 fathoms. Temperature of the bottom, 1·5° C. Sea bottom, globigerina ooze.

Station 310. January 10, 1876. Lat. 51° 30′ S., long. 74° 3′ W. Depth of the sea, 400 fathoms. Temperature of the bottom, 7·9° C. Sea bottom, mud.

Observations.—This species is distinguished from the foregoing by its much more slender form and shorter palpus. From *Colossendeis angusta*, G. O. Sars (Arch. f. Math. og Naturvid., vol. ii. p. 268, 1877), and other species, it may be easily distinguished by the dimensions of the joints of the palpus; for, as far as I know, *Colossendeis leptorhynchus* is the only species of this genus which has the fifth joint of the palpus considerably longer than the third. The form of the cephalic part of the cephalothorax, and the extremely small claws at the ends of the legs indicate, I believe, a near relationship between this species and *Colossendeis gigas*.

Colossendeis gigas-leptorhynchus.

A single specimen—unfortunately defective—dredged at Station 158, shows to a

---

[1] This, of course, refers to the animal preserved in spirits. I observed the same thing in living specimens of *Nymphon gracile*, Leach, and *Pallene brevirostris*, Johnston.

certain extent the characteristics of *Colossendeis gigas,* and in other respects those of *Colossendeis leptorhynchus.* I therefore wish to consider it as an intermediate form. Its dimensions are as follow :—

Length of the proboscis, . . .	26 mm.
Length of the trunk with the abdomen, .	18 ,,
Length of the palpus, . . .	33½ ,,
Length of the ovigerous leg, . .	?
Length of the leg, ...	137 ,,

As I consider that this specimen is not quite adult, I will not give a detailed description of it. I only wish to point out that the form of the proboscis, though a little more slender, quite agrees with that of *Colossendeis gigas,* whereas the relative length of the joints of the palpus, and that of the whole palpus, is the same as in *Colossendeis leptorhynchus,* viz., the third joint of the palpus is 8·5 mm., and the fifth 14 mm. The total length of the palpus is only three-quarters the length of the body.

The specimen is a young female, with the body almost smooth, and with legs only furnished with extremely small hairs.

Habitat.—This curious form was dredged at Station 158. March 7, 1874. Lat. 50° 1′ S., long. 123° 4′ E. Depth of the sea, 1800 fathoms. Bottom temperature, 0·3° C. Sea bottom, globigerina ooze.

Colossendeis robusta, n. sp. (Pl. IX. figs. 4, 5).

Diagnosis.—Proboscis club-shaped, shorter than the trunk. Body and legs rather stout, with large intervals between the lateral processes for the insertion of the legs. Palpi not very long, a great deal shorter than the body, with the third joint longer than the fifth. Legs not very slender, furnished with a claw, which is almost half as long as the second tarsal joint. Oculiferous tubercle conical, with four eyes.

Description.— ·

Length of the proboscis, ...	15 mm.
Length of the trunk with the abdomen, . . .	18 ,,
Length of the abdomen, ...	3·5 ,,
Length of the palpus, ...	25 ,,
Length of the ovigerous legs, ...	52 ,,
Length of the leg of the third pair, . . .	113 ,,

Only a single specimen of this beautiful and robust form was dredged. All the segments of the body are closely united, the cephalic part is comparatively short, and bears a conical, robust, not much elevated, oculiferous tubercle. This tubercle shows four distinct eyes, two of which are large and directed forwards, while the two small ones are directed backwards.

The intervals between the well-developed lateral processes are comparatively large. The proboscis is not quite half so long as the total length of the body. It is very stout, and shows a considerable swelling in the middle, and another at the extremity. The abdomen is small, its length about one-fifth the length of the trunk.

The palpi are comparatively stout, shorter than those of *Colossendeis gigas*, but longer than those of *Colossendeis leptorhynchus*. The first two joints are very small, the third is by far the longest of all the joints. The fourth is short again, the fifth only three-fourths as long as the third, the sixth to the tenth are nearly of the same length, the seventh, however, is a little longer. The joints are almost perfectly smooth, with the exception of some very small hairs visible only with the microscope.

The ovigerous legs have the first three joints short, the fourth and sixth of about the same length, and comparatively long. The fifth (the elbow-joint) is short. The four last joints are almost of the same length, becoming more slender from the first to the fourth. The claw is small and smooth; the place of the denticulate spines of the four last joints is filled up by short strong knobs of a conical or rounded shape. They are placed on excavations of the chitinous skin, and with the exception of the two rows on one side, are placed rather irregularly.

The legs are not very long. Those of the third pair are the longest. The first and the fourth leg of the right side are, in the Challenger specimen, quite rudimentary. In the full-grown leg the fourth and the sixth joint have the same length, whereas the fifth is a little shorter. Of the two tarsal joints the first is longer than the second. The claw is robust, about half as long as the second tarsal joint.

The only specimen brought home by the Challenger Expedition is a female. Its genital openings are not very large, and are found ventrally on the second joint of all the legs.

The animal is entirely smooth, and of a beautiful orange colour.

Habitat.—This species was dredged off Christmas Harbour, Kerguelen, 29th June 1874. Depth of the sea, 120 fathoms.

Observations.—This species cannot easily be confounded with any of the other species; it is distinguished by being stout, yet comparatively slender, by the form of the proboscis, and the presence of four distinct eyes. It is rather a shallow-water species.

Colossendeis megalonyx, n. sp. (Pl. IX. figs. 1–3).

Diagnosis.—Proboscis club-shaped, somewhat bent over to the ventral side. Cephalic part of the cephalothorax not distinctly separated. Oculiferous tubercle conical, with four eyes. Third joint of the palpus longer than fifth. Body and legs slender. Claws of the legs as long as the second tarsal joint.

Description.—

Length of the proboscis,	.	.	20 mm.	8 mm.	13·5 mm.	17·5 mm.
Length of the trunk,	.	.	10·5 ,,	8 ,,	8 ,,	10·5 ,,
Length of the abdomen,	.	.	2·5 ,,	2 ,,	1·5 ,,	2 ,,
Total length of the body,	.	.	33 ,,	...	...	...
Length of the palpus,	.	.	26 ,,	...	...	...
Length of the ovigerous leg,	.	.	39 ,,	...	... .	...
Length of the leg of the third pair,		.	97 ,,	...	...	...

The proboscis of this species is club-shaped, a little bent over to the ventral side ; longer than the trunk with the abdomen. The cephalothoracic segment is comparatively small, and not distinctly divided into a cephalic and a thoracic part. Between the lateral processes of the thoracic joints large intervals are to be observed. The oculiferous tubercle is large and high, conical, furnished with four eyes. Of these, two are very large and directed forwards, while the two directed backwards are a great deal smaller and in all probability are rudimentary. The abdomen is small, only 1-13th of the total length of the body.

The palpi are slender, and the lengths of their joints are not very characteristic. The third joint is much longer than the fifth, and of the three last joints the eighth is very small, the ninth and tenth considerably longer. In the ovigerous legs the fourth and sixth joints have nearly the same length, while the fifth is not quite half as long. The four last joints are small, and nearly equal in length, growing however a little shorter and more slender towards the tenth joint. The claw is very short.

The distribution and form of the denticulate spines is in this species very characteristic. There are two rows of comparatively long and flat spines, the margins of which show when greatly magnified very minute hairs ; in addition to these two rows a moderate number of smaller flattened spines are found scattered over that side of the four last joints which is bent inwards. I have figured these two rows and the irregularly placed spines in figure 3 on Plate IX.

The legs are not very long, nearly three times as long as the body in the adult specimen, while the younger specimens have them much shorter. The three first joints are very short, the fourth is the longest, viz., 23 mm. ; in a leg, the fifth joint of which measures 21 mm., the sixth is 18 mm. The two tarsal joints are together as long as the sixth joint. The claw is large, almost as long as the second tarsal joint.

The surface of the body and of the legs is almost entirely smooth. The palpi show only very small hairs, and on the ovigerous legs hardly any hairs are to be observed.

Of this species seven specimens were dredged. They are, I think, all young ones with the exception of one specimen which is a male. It shows genital pores ventrally on the second joint of all the legs.

Habitat.—This specimen was dredged at the following stations :—

Station 149. January 29, 1874. Off Christmas Harbour, Kerguelen. Depth of the sea, 120 fathoms.

Station 313. January 20, 1876. Lat. 52° 20′ S., long. 68° 0′ W. Depth of the sea, 55 fathoms. Temperature of the bottom, 8·8° C. Bottom of the sea, sand.

Station 314. January 21, 1876. Lat. 51° 36′ S., long. 65° 40′ W. Depth of the sea, 70 fathoms. Temperature of the bottom, 7·8° C. Bottom of the sea, sand.

Observations.—This species resembles *Colossendeis proboscidea*, Sab., sp. (Appendix to the Supplement of Captain Parry's Arctic Voyage, 1824, p. cexxvi.), in the form of the proboscis. That species, however, is a great deal stouter, and has a much larger body with comparatively short legs. Moreover, the body of *Colossendeis proboscidea* is rather disc-shaped, and by no means so slender as that of *Colossendeis megalonyx.* This species ranges from Kerguelen as far west as the east coast of Patagonia ; the three stations, however, at which it was found are nearly under the 50th parallel.

Colossendeis gracilis, n. sp. (Pl. IX. figs. 6–8 ; Pl. X. figs. 6, 7).

Diagnosis.—Body slender, with wide intervals between the lateral processes ; proboscis about as long as the trunk ; palpus once and a half as long as the proboscis, with the third joint longer than the fifth, the eighth joint extremely small, and the ninth joint laterally inserted on the front of the foregoing. Ovigerous legs about once and a half as long as the total length of the body. The claw of the legs longer than the second tarsal joint. Oculiferous tubercle conical, without eyes.

Description.—

Length of the proboscis,	.	.	.	6·5 mm.	5·8 mm.
Length of the trunk,	.	.	.	6 „	5·2 „
Length of the abdomen,	.	.	.	1·5 „	1 „
Total length of the body,	.	.	.	14 „	12 „
Length of the palpus,	.	.	.	9·5 „	9 „
Length of the ovigerous leg, .	.	.	.	22 „	19·5 „
Length of the leg of the third pair, .	.	.	.	51 „	40 „

The proboscis of this species is nearly as long as the trunk, and a little shorter than the trunk with the abdomen. It is a little swollen nearly in the middle, but in some specimens it is almost quite cylindrical. The cephalothoracic segment has the cephalic part, which is rather triangular, in some degree distinct. It bears anteriorly the very high conical oculiferous tubercle.

The lateral processes for the insertion of the legs are widely separated. The abdomen is not very short, and in one of the specimens is a little more swollen at the extremity than in the others.

Mandibles (Pl. X. fig. 6) are present in one of the specimens. It is a young male;

the largest specimen, however, of this species dredged. These mandibles are very long and slender, three-jointed, little shorter than the proboscis. The third joint terminates in a pair of small and slender pincers.

Of the palpi the third joint is longer than the fifth. Of the last three joints the first is extremely small and broad (Pl. IX. fig. 7), and the ninth is inserted, not into the middle of the front part of that joint, but quite laterally. This ninth joint itself is cylindrical, and more than twice as long as the foregoing, whereas the tenth is a little more slender and still longer than the ninth joint.

The joints of the comparatively long ovigerous legs have characteristic proportions in all the species of the genus *Colossendeis*. The claw at the end of the tenth joint is very small. The denticulate spines on the last four joints are not numerous, the number of rows not exceeding four. Of these two are very regular, with the spines placed close to one another, but those of the two other rows are much more irregularly scattered over the remaining part of the inner surface of the joint (Pl. IX. fig. 8 ; Pl. X. fig. 7).

The lengths of the several legs of one and the same specimen only show slight differences. The third leg is, in the three specimens I have examined, the longest, and measures nearly three and a half times the length of the body. The claw is very long, still longer than the second tarsal joint. Genital pores are present on the second joint of every leg. As to the sex of the three specimens dredged during the voyage of H.M.S. Challenger, I have only been able to ascertain that the specimen with mandibles is a male.

The body and legs are almost entirely smooth ; small hairs are only to be observed on the last joints of the palpi.

Habitat.—This species was dredged between the Cape of Good Hope and Kerguelen Island.

Station 146. December 29, 1873. Lat. 46° 46′ S., long. 45° 31′ E. Depth of the sea, 1375 fathoms. Temperature of the bottom, 1·5° C. Bottom of the sea, globigerina ooze.

Station 147. December 30, 1873. Lat. 46° 16′ S., long. 48° 27′ E. Depth, 1600 fathoms. Bottom temperature, 0·8° C. Bottom of the sea, globigerina ooze.

Observations.—To this species *Colossendeis media* and *Colossendeis brevipes*, which I shall describe further on, are closely allied. The occurrence of mandibles in one of the specimens of this species is very interesting. Although larvæ of any species of *Colossendeis* have never been observed, they doubtless are furnished with three pairs of cephalic appendages. Of these the adult animal has always lost the first pair ; and whenever, as in the case in question, this first pair is observed in the adult animal, it must be considered as a case of atavism, showing that the loss of mandibles in the adult animal has been comparatively recent.

Colossendeis media, n. sp. (Pl. X. figs. 10, 11).

Diagnosis.—Palpi more than once and a half the length of the proboscis; ninth joint not attached laterally to the eighth; together they are as long as the tenth joint. Denticulate spines of the four last joints of the ovigerous legs in five distinct rows. Legs not quite three times as long as the body. Otherwise this species resembles the foregoing.

Description.—

Length of the proboscis, ...	7½ mm.
Length of the trunk, ...	6½ „
Length of the abdomen, ...	1½ „
Total length of the body, ...	15¼ „
Length of the palpus, ...	12 „
Length of the ovigerous leg, ...	20 „
Length of the leg of the third pair, ...	45½ „

This species is nearly of the same shape and the several parts show almost the same proportions as in *Colossendeis gracilis*, Hoek. The only differences are the following :—

1. The palpi are longer, and of the three last joints the very small eighth joint does not bear the following joint laterally, but quite in front (Pl. X. fig. 10).

2. The denticulate spines of the ovigerous legs (of the four last joints) are placed in five distinct rows. They are small with the exception of those of the outermost row, which are a great deal larger. These spines have the flattened shape of those of other species of *Colossendeis*. Those which are not broken, show on the margin very small hairs.

3. The legs are not quite thrice as long as the body. The fourth joint is twice as long as the sixth. The claw is longer than the second tarsal joint. The two tarsal joints are about the same length, together they are as long as the body.

The body and legs are almost entirely smooth; when seen with the microscope very short hairs are to be observed; those on the palpi are a little stronger. On the second joint, on the ventral surface, of every leg very small genital pores are to be observed. However, as the two specimens of this species probably are not quite adult, I could not determine to which sex they belong.

Habitat.—This species was dredged at Station 298. November 17, 1875. Lat. 34° 7′ S., long. 73° 56′ W. Depth of the sea, 2225 fathoms. Bottom temperature, 1·3° C. Bottom of the sea, grey mud.

Observations.—This species is nearly allied to *Colossendeis gracilis*. However, as the differences I have pointed out are present equally in both specimens, I cannot consider them identical with that species.

Colossendeis brevipes, n. sp. (Pl. X. figs. 8, 9).

Diagnosis.—Lateral processes not very widely separated. Palpus less than once and a half as long as the proboscis, with the three last joints bent over like a hook. Ovigerous legs with the denticulate spines of the four last joints numerous and comparatively short Length of the legs unequal, short. Otherwise this species resembles *Colossendeis gracilis*, Hoek.

Description.—

Length of the proboscis,	.	9 mm.	8·5 mm.
Length of the trunk,	.	7·5 ,,	7·5 ,,
Length of the abdomen,	.	2·5 ,,	2 ,,
Total length of the body,	.	19 ,,	18 ,,
Length of the palpus,	.	12 ,,	11·5 ,,
Length of the ovigerous leg, .	.	29 ,,	25 ,,
Length of the leg of the first pair,	.	50 ,,	54 ,,
Length of the leg of the second pair, .	.	62 ,,	77 ,,
Length of the leg of the third pair,	.	53 ,,	62 ,,
Length of the leg of the fourth pair, .	.	45 ,,	52·5 ,,

This true deep-sea species is also nearly allied to *Colossendeis gracilis*. The first time I examined it I was struck with the shortness of the legs, especially in the largest specimen; but as it is possible that this length varies greatly under different circumstances—as it certainly does with age—it is necessary to use the utmost care in judging of this characteristic mark. The body, proboscis, trunk, and abdomen are as in *Colossendeis gracilis*, the only difference being that the lateral processes are not quite so widely separated.

The palpus is short, only one-fourth as long as the proboscis. The last three joints are extremely short, shorter than the seventh joint ; they are bent like a hook, the tenth joint hanging down parallel with the seventh joint (fig. 8).

The ovigerous legs are comparatively long, 29 mm. in a specimen of 19 mm. The last four joints are small and furnished with a rudimentary claw, the length of which is nearly one-fifth the length of the tenth joint. A drawing of the denticulate spines on one of the four last joints of the ovigerous leg is given in fig. 9. The spines are short and flat and do not show small hairs at the periphery. There are two regular rows, the outermost of which has the stoutest and longest spines, whereas the irregularly placed spines on the other side decrease in size. The length of the legs is very unequal. The specimen with the shortest legs is a full-grown female. The claw at the end of the eighth joint is very long, longer than that joint. The body and legs are almost entirely smooth ; there are small hairs on the last joints of the palpi.

The genital pores of both specimens, the largest of which is a female, are found ventrally on the second joint of all the legs.

Habitat.—This species was dredged from the greatest depth at which a Pycnogonid has been found, viz., 2650 fathoms.

Station 325. March 2, 1876. Lat. 36° 44′ S., long. 46° 16′ W. Depth of the sea, 2650 fathoms. Bottom temperature, 70·4° C. Bottom of the sea, grey mud.

Observations.—Whether I am right or not in considering the specimens collected at Stations 146 and 147 (*Colossendeis gracilis*, Hoek), Station 298 (*Colossendeis media*, Hoek), and Station 325 (*Colossendeis brevipes*), as three different species can only be ascertained by examining a larger number of specimens than are at my disposal. I can only point out here the great affinity of these different specimens. However, I may be permitted to state here, that whenever I speak of a specimen as adult, the statement is based on the microscopical examination of transverse sections of the fourth joint of one of the legs.

Colossendeis minuta, n. sp. (Pl. X. figs. 12–14).

Diagnosis.—Proboscis cylindrical, once and a half the length of the trunk with the abdomen. Lateral processes not widely separated. Palpus once and a half as long as the proboscis, with the last three joints of about the same length. Legs extremely slender, about four and a half times the length of the body. Claw of the leg half as long as the second tarsal joint.

Description.—

Length of the proboscis,	.	.	.	.	.	.	4·8 mm.
Length of the trunk,	.	.	.	.	.	.	2·7 „
Length of the abdomen,	.	.	.	.	.	.	0·5 „
Total length of the body,	.	.	.	.	.	.	8 „
Length of the palpus,	.	.	.	.	.	.	7·5 „
Length of the ovigerous leg,	.	.	.	.	.	.	14 „
Length of the leg of the third pair,	.	.	.	.	.	35 „

Only a single specimen of this species was dredged. It is a small animal with a comparatively long proboscis and very slender legs. The body is not extremely slender, there being only small intervals between the lateral processes for the insertion of the legs. The cephalothoracic segment is small and not distinctly divided into a cephalic and a thoracic part. Quite near the front it is furnished with a blunt oculiferous tubercle without eyes.

The proboscis is nearly cylindrical, it is a little swollen in the middle, and the mouth is small and triangular. The abdomen is small; it is connected with the last thoracic joint by a true articulation.

The palpus is slender, but not extremely long, being only once and a fourth as long as the proboscis. The third joint is only a little longer than the fifth. I have given a drawing of joints six to ten in fig. 13. The last three joints together are longer than the seventh; they are of about the same length, and comparatively slender. Beginning with the fifth, all the joints are furnished with short but strong hairs, which have the same shape as those on the legs and ovigerous legs.

The ovigerous legs are of the ordinary form. Only the middle of that part of the surface of the four last joints which is turned inwards, is furnished with denticulate spines. Their form is very peculiar, being flattened and much broader towards the extremity (Pl. X. fig. 14). They are not very numerous and are placed in three rows only, while every row contains about six of them. The margin of the spines is smooth. The claw of the ovigerous leg is small, and the length of the several joints is as in the other species of the same genus.

The legs are extremely long and slender. With the exception of some small but strong hairs, like those of the palpus, they are smooth. The first three joints are small; of the three following the first two (the fourth and fifth) are nearly of the same length, whereas the sixth is considerably shorter, but much longer than the last two joints together. Of these the first (the first tarsal joint) is about once and a half as long as the second. The claws at the ends of the legs are about half as long as the second tarsal joint.

About the sex of the only specimen collected I am unable to form an opinion. Perhaps it is not yet quite adult: extremely small genital pores are to be observed ventrally on the second joint of every leg.

Habitat.—This interesting little *Colossendeis* was dredged by the Challenger at Station 50. May 21, 1873. Lat. 42° 8′ N., long. 63° 39′ W. Depth of the sea, 1250 fathoms. Temperature of the bottom, 2·8° C. Bottom of the sea, grey ooze.

Observations.—This curious form is intermediate, I believe, between the short-nailed *Colossendeis leptorhynchus*, with its extremely long proboscis, and the long-nailed *Colossendeis gracilis*, with the comparatively short proboscis. The difference in length between the third and the fifth joint,—in *Colossendeis leptorhynchus* the fifth joint is the longest, and in *Colossendeis gracilis* it is the third joint which has the advantage—has almost disappeared in *Colossendeis minuta*.

The species was dredged about two degrees south of Halifax; though the Pycnogonids of the neighbourhood of the coast of New England are comparatively well known, the occurrence of a species of *Colossendeis* in those regions has not been recorded as yet. And this, no doubt, is due to the circumstance that the species in question inhabits deep water, while hitherto only the shallow water inhabitants have been carefully investigated.

Discoarachne, n. gen.

Diagnosis.—Proboscis stout, cylindrical, tapering towards the extremity. Mandibles wanting. Palpi five-jointed. Ovigerous legs not furnished with denticulate spines, ten-jointed.

Discoarachne brevipes, n. sp. (Pl. VII. figs. 8–12).
Diagnosis.—Body disciform, consisting of the true body without any segmentation

and the closely united lateral processes. Legs short with auxiliary claws. Body and legs smooth, furnished only with very small hairs.

Description.—

Length of the proboscis,	.	.	.	.	.	.	2½ mm.
Length of the abdomen,	.	.	.	.	.	.	1¼ „
Total length of the body,	.	.	.	.	.	.	4¼ „
Length of the palpi,	.	.	.	.	.	.	1½ „
Length of the ovigerous legs,	.	.	.	.	.	.	2⅜ „
Length of the legs, .	.	.	.	.	.	.	7 „

The cephalothoracic joint is closely connected with the three other thoracic joints ; the lateral processes are short and somewhat triangular, forming in connection with the body a disciform surface, from which the legs radiate in different directions. The proboscis is comparatively large, the form cylindrical, tapering towards the extremity. The oculiferous tubercle, which is small, not very elevated, and furnished with four distinct eyes, is placed dorsally on the middle of that part of the body, which corresponds with the cephalothoracic segment.

The palpi are small, measuring about two-thirds the length of the proboscis. The first and second joints are very short, the third joint is the longest of all, the fourth again small, the fifth more than twice as long as the fourth, and comparatively slender. The first two joints are smooth, the third joint is furnished up to the outer margin with three long spines, and on its inside with three smaller spines, which are curved ; the ventrally directed surface of the same joint, which contains a large gland, hereafter to be described, has, near the middle, four extremely small spines. The fourth joint is nearly smooth, the fifth is armed with very large and numerous spines near the extremity.

The first joint of the ovigerous legs is very small, the second and the third are a little longer, the fourth and the fifth are the longest, the sixth is short again, the seventh, eighth, ninth, and tenth are about the same length, and gradually diminish slightly in breadth. The first five joints are nearly smooth, towards the distal extremity the sixth shows some short and not very strong spines, whereas the four following joints are furnished in the same place with much stronger and slightly curved spines. Towards the end of the tenth joint there are a considerable number of these spines, while a claw and true denticulate spines are totally wanting.

The first three joints of the legs are small ; of the three following, the middle joint is the longest. The first tarsal joint is extremely short, the second comparatively long, and very slender in comparison with the other joints of the legs. The claw is about one-third the length of the second tarsal joint. The auxiliary claws are comparatively long, two-thirds the length of the claw. The hairs on the different joints of the legs are small ; the distal extremity of these joints is, as a rule, furnished with a row of longer and stronger spines, especially at the end of the sixth joint. The first tarsal joint has numerous small hairs on the inner margin; the second is armed with three very strong and broad spines on the inner margin near the base, while longer and much more slender spines are present on the outer margin.

A single specimen of this species was collected during the cruise of H.M.S. Challenger. It is a female with eggs in all stages of ovarian development, enclosed in the fourth joint of the leg. The genital pores I could not distinctly observe, but no doubt they are placed ventrally near the distal extremity of the second joint of the leg.

Habitat.—This species was found at Seapoint near Cape Town in November 1873.

Observations.—The genus *Discoarachne* is in all probability nearly allied to *Endeis*, Philippi. But as that genus is difficult to make out from the insufficient description of Philippi, I propose a new genus for the species in question, because Philippi's form had eight-jointed palpi, whereas the Challenger form, which must be a full-grown animal, has only five-jointed palpi. This species with its short legs and highly concentrated body, is a true littoral form.

Pallene, Johnston.

Pallene australiensis, n. sp. (Pl. XI. figs. 1–7).

Diagnosis.—Body extremely slender. Cephalic part of the cephalothoracic segment distinctly separated from the thoracic part by means of a true articulation. Proboscis short, inserted ventrally. Ovigerous legs with true denticulate spines, and a long denticulate claw. Legs without auxiliary claws.

Description.—

Length of the proboscis,	.	.	.	.	.	.	1·5 mm.
Total length of the body,·	.	.	.	.	.	.	6·5 ,,
Length of the ovigerous legs,	.	.	.	.	.	7 ,,
Length of the leg of the third pair,	.	.	.	.	28 ,,

The body is extremely slender; the cephalic part of the cephalothoracic segment (neck) is much swollen at the tip for the insertion of the mandibles. A true articulation divides this cephalic part from the thoracic part. The rather short oculiferous tubercle is situated about the front of the thoracic part.

The intervals between the lateral processes, at the extremities of which the legs are found, are very large. The abdomen is rudimentary and directed a little upwards. The surface of the body and of the lateral processes is smooth. The proboscis is short and inserted rather ventrally. It is constricted beyond the middle, and is much wider at the extremity, where the very small triangular mouth is observed.

The mandibles are short and robust. The first joint reaches almost as far as the end of the proboscis, the second joint is considerably swollen, and terminates in a pair of short pincers. These are furnished with a single very small tooth towards the extremity.

The ovigerous legs are inserted ventrally close to the lateral process of the first pair of legs. Of the first three joints, which are small, the third is the longest; the fourth is a great deal longer, and the fifth is the longest of all, in the males it has a strong knob near the extremity ; the sixth joint is almost as long as the fourth. The four last joints show no great difference ; from the seventh to the tenth they gradually diminish in length and in breadth. The claw is as long as the tenth joint, and on the inside is furnished with a row of small spines. The hairs on the joints of the ovigerous legs are not very strong. I have figured the denticulated spines in Pl. XI. fig. 5.

The legs are very long. The second joint is thrice as long as the first and as the third. In the female it is considerably swollen towards the extremity, where the genital opening is observed on the ventral surface. Of the three following joints the fourth and the fifth are nearly of the same length, while the sixth is almost once and a half as long. The first tarsal joint is extremely short, and the second about eight times as long. The claw is half as long as the second tarsal joint. The joints of the leg are furnished with extremely long and slender spines ; two are placed at the distal extremity of the first joint, two in the middle of the second joint, and two others towards the extremity of that joint ; a row of slightly shorter ones is observed on the dorsal surface towards the extremity of the third joint ; the fourth, fifth, and sixth joints are also furnished with some of these hairs placed at irregular distances from one another ; towards the extremity of these joints a certain number of these hairs is invariably observed. The first tarsal joint (fig. 6) shows only a few very short spines, whereas the second tarsal joint towards the side opposite to the claw is furnished with a complete row of very stout spines. Most of these spines, like the greater part of those placed on the other joints, are not quite smooth, but are furnished on one side with sharp short triangular hooks not unlike the teeth of a saw.

Of the three specimens of this species, one is a female, and two are males. I believe they are all full-grown animals. One of the males shows the remains of egg-packets adhering to the ovigerous legs. The genital openings of the females are much larger than those of the males. Both sexes have them placed ventrally on the second joint of all the legs.

Habitat.--This beautiful species was dredged in the neighbourhood of Melbourne.

Station 162. April 2, 1874. Off East Moncoeur Island, Bass Strait. Depth of the sea, 38 to 40 fathoms. Bottom of the sea, sand.

Also between Melbourne and Sidney at Station 163. April 4, 1874. Lat. 36° 56′ S., long. 150° 30′ E. Trawled in 120 fathoms. Off Twofold Bay.

Observations.—Perhaps this species is nearly allied to *Pallene chiragra*, Milne-Edwards, Histoire naturelle des Crustacés, tom. iii. p. 535. He gives the following description of this species :—" Corps très-grêle ; tête courte mais cylindrique. Second article des pates-mâchoires très-renflé, et premier article du thorax extrêmement allongé. Pates environ cinq fois aussi longues que le corps, sans crochets accessoires au bout. Pates accessoires de la femelle de dix articles." However, as in this description the very distinct spines on the legs have not been mentioned, which, if present, would certainly have been seen by Milne-Edwards, I think it probable that his *Pallene chiragra* is a nearly allied but distinct species. Jarvis Bay, New Holland, where Milne-Edwards's species was collected, is not far from Station 163.

Pallene lævis, n. sp. (Pl. XI. figs. 8–12).

Diagnosis.—Body robust, lateral processes scarcely separated. Body and legs smooth. Proboscis short, conical; inserted about the front of the cephalothoracic segment. Ovigerous legs with denticulate spines and a long claw. Legs without auxiliary claws.

Description.—

Length of the proboscis,	1 .mm.
Length of the trunk,	2·66 „
Length of the abdomen,	5·0 „
Total length of the body,	4 „
Length of the ovigerous leg,	6 „
Length of the leg of the third pair,	21·5 „

The body of this species is robust. The cephalothoracic segment is of a curious shape : it is considerably swollen at the front, where it bears the proboscis and the mandibles; it is constricted in the middle, thus forming a sort of short neck, and it is much wider again at the back, where it bears dorsally the oculiferous tubercle, and ventrally the short lateral processes for the insertion of the ovigerous legs. The oculiferous tubercle, with two larger eyes directed forwards and two smaller ones backwards, is situated almost exactly above the insertion of the ovigerous legs. The lateral processes for the insertion of the legs are comparatively long; the abdomen is short and stout.

The proboscis is short and conical, and has a very small mouth at the extremity. The mandibles are rather stout. The basal joint is constricted at the base, and indistinctly divided into two joints, it is nearly as long as the proboscis ; the second joint is placed at right angles with the basal joint, and is considerably swollen and stout. At the extremity it is furnished with two claws, one straight, pointed and immovable, the other curved and movable, but also pointed. The inner surface of these claws is smooth, but there is a blunt point in the middle of the movable claw.

The ovigerous legs of the female specimen (the only one dredged) are not very strong. The first three joints are small, the fourth and the fifth are the longest, nearly of the same length and a little curved. The sixth joint is not quite half as long as the fifth. The four last joints are but little shorter than the sixth. This claw is comparatively strong: it is not denticulated on the inner surface but slightly serrated. The joints of the ovigerous legs are almost quite smooth. The shape of the denticulate spines is very curious. They are broad and flat, have two or three stronger teeth on each side near the base, and extremely fine teeth all over the rest.

The very smooth legs are comparatively long. The first and the third joints are short and almost of equal length, the second joint is more than twice as long, having a large oval genital opening at the extremity on the ventral surface. The fourth joint, containing the ovary, is considerably swollen and very long; the fifth is only a little shorter, the sixth, on the contrary, is a little longer. The two tarsal joints are very small, together about one-fifth of the length of the sixth joint. Almost every joint describes a feeble but characteristic curve; especially the second, the fourth and the eighth joint. The first tarsal joint is extremely small, and is furnished with a large number of hairs and an isolated stronger spine, the second tarsal joint also shows a number of hairs and four stronger spines opposite to the claw. The claw is strong and very long, considerably curved, and without auxiliary claws.

The only specimen of this species dredged during the cruise of H.M.S. Challenger is a female, which, I think, is a full-grown one.

Habitat.—This species, along with a specimen of *Pallene australiensis*, Hoek, was collected at Station 162. April 2, 1874. Off East Moncœur Island, Bass Strait. Depth of the sea, 38 to 40 fathoms. Bottom of the sea, sand.

Observations.—This very characteristic species may be easily recognised among the different species of *Pallene* by the form of the proboscis and cephalothoracic segment, by the shape of the denticulate spines of the ovigerous legs, by the presence of a claw at the end of the ovigerous leg, and finally by the absence of auxiliary claws at the end of the legs.

Pallene languida, n. sp. (Pl. XII. figs. 1–5).

Diagnosis.—Body highly concentrated, rather disciform. Proboscis conical, cephalothoracic segment comparatively long. Ovigerous legs with denticulated spines, but without a claw. Legs without auxiliary claws. Oculiferous tubercle conical, elevated. Rudiments of palpi in the form of knobs.

Description.—

Length of the proboscis,	.	.	.	.	.	0·45 mm.
Length of the trunk,	.	.	.	.	.	1·35 „
Total length of the body,	.	.	.	.	.	1·8 „
Length of the ovigerous leg,	.	.	.	.	.	2·3 „
Length of the leg of the third pair,	.	.	·.	.	.	5·1 „

Of this curious species, unfortunately, only a single specimen was collected, and this specimen had, moreover, suffered much from the alcohol; it is visibly crumpled, especially on the cephalothoracic segment, as seen from the dorsal side; the other segments have also suffered in a less degree, which makes it very difficult to judge of the original form.

The proboscis is small, obtusely conical; the mouth is small, as in the other species of this genus. The cephalothoracic segment is considerably swollen anteriorly, where it bears the proboscis and the mandibles; in the middle it is constricted so as to form a neck, and posteriorly it becomes wider again. This wider part (the thoracic part of the cephalothorax) bears the conical oculiferous tubercle, which shows only rudimentary eyes. The cephalothoracic segment seems, like the two following segments, to get considerably wider posteriorly on the dorsal surface, thus forming large folds in the breadth above the insertion of the first three pairs of legs. The fourth thoracic segment is quite flat; the lateral processes for the insertion of the legs are extremely short. The abdomen is very small but comparatively broad.

The surface of the body is smooth; on the mandibles, the ovigerous legs, and the legs, numerous not very long but spiny hairs are observed. The first joint of the mandibles reaches as far as the tip of the proboscis. The second joint is not very strong, is furnished with slender pincers, and is armed on the inner side with four short teeth. Rudiments of palpi are implanted ventrally near the base of the mandibles; they are only one-jointed knobs.

The ovigerous legs are short, little longer than the body. The first three joints are short; the two following joints are much longer; the fifth is the longest and is armed near the distal extremity with a small knob; the sixth joint is short. Of the four following joints the first is the longest, the last three are nearly of the same length. There is no claw at the end of the tenth joint. The denticulate spines of one and the same joint are by no means all of the same shape, those placed near the preceding joint are a great deal smaller than those in the middle of the row, whereas those placed at the end of the row are the largest of all, and are furnished at the base with three strong teeth of which the third especially is very large (fig. 5).

The length of the legs is almost three times the length of the body. The first three joints are small; the three following are much longer, but not very slender. The seventh joint is extremely short; the second tarsal joint is nearly six times as long. Besides the hairy spines on all the joints, which, as far as I could ascertain, are scattered rather irregularly over the surface, the last joint has, on the inner side, a row of short and comparatively strong spines. The claw at the end of the leg is long and stout. Auxiliary claws are wanting.

Of this species only a single specimen was obtained during the cruise of the Challenger, and, judging from the knobs at the end of the fifth joint of the ovigerous leg, I consider it to be a male.

Habitat.—*Pallene languida* was dredged at

Station 161. April 1, 1874. Off the entrance to Port Philip (Melbourne, Australia). Depth of the sea, 38 fathoms. Bottom of the sea, sand.

Observations.—This species is in all probability nearly allied to *Pallene longiceps*, Böhm (Sitzungsberichte der Gesellsch. Naturf. Freunde in Berlin, 1879, p. 59). However, as no figure of Böhm's species has ever been published, and as the description of it cannot be entirely applied to my specimen, I thought it safer to consider, and to describe this as a new species.

Pallene longiceps, Böhm, has rudimentary two-jointed palpi, and a short and blunt oculiferous tubercle. Moreover, the form of the denticulated spines of the ovigerous legs of the present species is different from the form described by Böhm for his *Pallene longiceps*. The latter species is from Japan, whereas my *Pallene languida* was obtained in the vicinity of Melbourne.

Phoxichilidium, Milne-Edwards.

Phoxichilidium fluminense, Kröyer (Pl. XIV. figs. 1–4).

Phoxichilidium fluminense, Kröyer, Bidrag til Kundskab, &c., Naturh. Tidskr. Ny Raekke, vol. i. p. 124, 1845, Tab. i. fig. 1*a–f.*

Pallene fluminensis, Kröyer (sp.), Böhm, Pycnogoniden des Museums zu Berlin, Monatsbericht der k. A. der Wissensch. zu Berlin, Februar 1879, p. 180, Taf. i. fig. 4–4*f.*

This species has been described and figured by Kröyer (*loc. cit.*), and again by Böhm (*loc. cit.*). It may not, however, be considered superfluous to publish new figures ; those of Kröyer are in general highly characteristic, but they are, as regards the details, not very exact ; from Böhm's figure, which has been drawn on much too small a scale, nobody, I think, would recognise the species.

The description given by both authors is nearly correct. The basal joint of the mandible reaches farther than the tip of the proboscis ; it shows dorsally a little beyond the middle a slight angle, furnished with a row of stronger hairs : therefore the joint seen from the dorsal surface seems to be divided into two. Rudiments of palpi are present in the form of rounded knobs on both sides of the cephalic segment. The ovigerous legs of the full-grown animal are ten-jointed ; I have figured joints six to ten in fig. 3. The sixth joint shows a wreath of short strong spines immediately before the articulation with the seventh joint. The seventh to the tenth joints are armed with curved spines and strong hairs, but no denticulate spines at all are present. These last five joints of the ovigerous leg are very curiously bent in the form of an S, as has been correctly observed and drawn by Kröyer. The legs are comparatively stout. The only specimen of this species brought home by the Challenger is a male, with rounded, rather large genital openings, which, as far as I could ascertain, are present only on the ventral surface of the second joint of the two hindermost legs. The fourth joint of the leg is more than twice as long as the second, and not inconsiderably swollen in the Challenger specimen ; ventrally a little in front of the middle it is furnished with a distinct and comparatively strong tubular process, which in all probability communicates with a gland situated in the interior of the joint. This tubular process has been observed neither by Kröyer nor by Böhm. Probably it occurs only in the male sex. The fifth joint is as long as the fourth, the sixth a little longer. The first tarsal joint is short, the second comparatively long, armed with a claw and two long auxiliary claws. Joints one to four have only a few hairs, while joints five to eight are covered by numerous, and for the most part, comparatively long and stout hairs.

Habitat.—This species, according to Kröyer, is found off Rio de Janeiro, whereas Böhm describes specimens collected by the German man-of-war, the "Gazelle," in the Straits of Magellan, and on the Patagonian coast, at a depth of 30 to 42 fathoms. One specimen was dredged by the Challenger off Bahia. Depth, from 7 to 20 fathoms.

Observations.—Böhm considers·this species a *Pallene*. I think, however, there can be no doubt that it is a true *Phoxichilidium* in the sense of Kröyer. To take the number of joints of the ovigerous legs as a decisive proof in this matter is, I think, not safe. The ovigerous legs of the *Pallene* (*Pallene lappa*, Böhm) which Böhm examined, were only seven-jointed, and for that reason alone the specimen cannot be considered as a *Phoxichilidium*. Of much greater value, I think, is the form of the ovigerous leg itself, the form of the last joints, of the spines with which they are furnished, &c., also the whole form of the body, the manner in which the cephalic part of the cephalothoracic segment overhangs the proboscis, &c.

Another question is, of course, whether it would not be convenient to class as a separate genus those forms of *Phoxichilidium* which have ten-jointed ovigerous legs, probably always present in both sexes. But before taking this step, the different forms ought to be better known, and for this a close study of full-grown specimens of both sexes is necessary. The genus *Anoplodactylus* of E. B. Wilson cannot be accepted, because neither the presence or absence of auxiliary claws, nor the fact of the ovigerous legs being five- or six-jointed, has any real importance.

Phoxichilidium insigne, n. sp. (Pl. XIV. figs. 5–7).

Diagnosis.—Body slender, with large intervals between the lateral processes. Proboscis cylindrical, inserted ventrally far posteriorly between the two ovigerous legs. Mandibles two-jointed, the first joint bearing the second laterally. Ovigerous legs six-jointed. No auxiliary claws. Legs and mandibles furnished with large conical knobs.

Description.—

Length of the proboscis,	.	.	.	.	.	2 mm.
Total length of the body,	.	.	.	.	.	6 „
Length of the abdomen,	.	.	.	.	.	0·5 „
Length of the ovigerous leg,	.	.	.	.	4·2 „
Length of the leg of the first pair,	.	.		.	.	19 „

Of this most curious form, unfortunately, only a single specimen—and that much mutilated, with only five legs—was collected during the voyage of the Challenger. This specimen is, I think, a full-grown male. The body is extremely slender, with very large intervals between the lateral processes for the insertion of the legs. The cephalo-thoracic segment is rather short, about twice as long as the first true thoracic segment; the second thoracic segment is a little longer than the first, and the third is, again, a little shorter. The proboscis is long, inserted ventrally, far back between the two ovigerous legs. It shows a little swelling at its base, in the middle, and again at the extremity. The mouth is small and triangular. The abdomen is short and directed somewhat upwards.

At the base of the mandibles the front part of the cephalothoracic segment is for a short distance cloven in the middle. Immediately behind this cleft the blunt oculiferous tubercle, with its four dark eyes, is placed. The basal joints of the mandibles diverge considerably, and extend beyond the front of the proboscis. The end of this basal joint, which is directed forwards, terminates in a pointed appendage, and laterally, under-neath the end of this appendage, the short second joint is attached. This has the form of a bird's head with the small pincers as jaws. The inner surface of these pincers is smooth. The first joint towards its extremity and the whole surface of the second joint are covered with numerous long hairs.

The ovigerous legs are inserted close to the base of the proboscis; they are six-jointed; the first joint is small, the second about half as long as the proboscis, the third nearly as long as the proboscis, the fourth half as long as the second, the fifth a little shorter than the fourth, and the sixth extremely small. All the joints are furnished with numerous small hairs; those on the two last joints are a little longer, but still extremely slender.

The first joint of the legs is small, nearly as long as the lateral process, the second joint is more than twice as long as the first, the third joint nearly half as long as the second, the fourth is as long as the trunk with the abdomen, the fifth is but little shorter, the sixth is as long as the whole length of the body, the seventh is short, and the eighth nearly as long as the second joint. The part of the eighth joint facing the strong claw has a distinct shoulder, furnished with spines and small hairs. The first joint of the leg bears at the distal extremity, on both sides, a strong conical protuberance; the second bears a still larger one ventrally, and another at the distal extremity; the third, too, is furnished with one. Besides three strong protuberances at the distal extremity, the longest of which is placed between the two others, the fourth joint has three other protuberances placed laterally on the joint; one of these is placed in the middle, the two others on the other side at equal distances from the middle one. The protuberances at the extremity of the joint are much larger than the others found on the joint. The latter have, moreover, a slender spine at the top. The armature of the fifth joint is nearly the same as that of the fourth. The sixth joint is furnished with numerous small protuberances, bearing slender spines at the top. Distinct hairs are seen on all the joints; towards the fourth joint they grow stronger and denser. The side of the last joint facing the claw when it is closed is furnished, in addition to numerous slender spines, with a row of curiously-shaped teeth (see fig. 7 of Pl. XIV.). The last joint terminates in a strong protuberance, like those placed at the extremities of the other joints of the leg. The claw is long and stout; auxiliary claws are wanting.

I think the only specimen of this species dredged by H.M.S. Challenger is a male. I could not, however, ascertain the sex without injuring the specimen, and I can only state my supposition. It is based on the fact that there are species of *Phoxichilidium* in which six-jointed ovigerous legs are present only in the male, and also on the presence of dermal glands in the fourth joint of the leg.

Habitat.—This curiously-shaped Pycnogonid was dredged off Bahia at a depth of 7 to 20 fathoms.

Observations.—I think this species of *Phoxichilidium* is a near relation of the European shore and shallow-water forms of the same genus (*Phoxichilidium femoratum, P. virescens,* &c.); from these it can be easily distinguished by the extremely characteristic protuberances on the legs, mandibles, &c.

Phoxichilidium patagonicum, n. sp. (Pl. XII. figs. 6–9).

Diagnosis.—Body robust, lateral processes scarcely separated. Basal joint of the mandibles indistinctly divided into two joints, second joint short, with small pincers. Palpi represented by large rounded knobs. Ovigerous legs ten-jointed, without claws, present in both sexes; auxiliary claws on the legs. Abdomen long.

Description.—Of this species a full-grown female and two small specimens, about whose sex I do not feel quite sure, were dredged by H.M.S. Challenger. For the description I have made use of the full-grown female.

Length of the proboscis,	6 mm.
Total length of the body,	16 „
Length of the abdomen,	5 „
Length of the ovigerous legs,	11 „
Length of the leg of the third pair,	57 „

The body of this species is almost entirely smooth; the cephalothoracic segment, which is not quite so long as the abdomen, bears the oculiferous tubercle quite at the front. The two following thoracic segments are together not quite so long as the cephalothoracic segment. The last segment is very small, it bears a long abdomen directed upwards. The length of the lateral processes is very considerable.

The proboscis is ventrally inserted, its base is found considerably behind the front margin of the cephalothoracic segment. It is comparatively long, and its shape is cylindrical; the extremity is rounded, with a small triangular mouth.

The mandibles are inserted close to each other and have a very long basal joint, which, seen from the dorsal side about the middle, shows a distinct articulation. This basal joint is considerably swollen at the extremity ; the second joint is directed towards the ventral side, while its pincers are bent laterally, so as to be directed towards those of the other mandible. These pincers are smooth and extremely short. While the basal joint of the mandibles is nearly smooth, the second joint is furnished with numerous and strong hairs, which are a little stronger still at the base of the pincers.

The palpi are represented by large rounded tubercles, placed at both sides of the base of the proboscis.

The ovigerous legs are comparatively short (at least in the female). The first joint is very small, the second is elongated and not quite three times as long as the first, the third is again short, the fourth and fifth joints are longer, the sixth is only two-thirds the length of the fifth, and the last four joints are small. They are figured on Plate XII. fig. 8, and are covered with long spiny hairs. There is no claw at the end of the tenth joint.

The first joint of the legs is small, the second is more than twice as long, and becomes considerably thicker towards the extremity, the third joint is only a little longer than the first, the three following are about the same length ; the fifth joint, however, is the smallest, the sixth the longest. This joint in the second leg of the right side describes a slight curve, which at the convex side is surmounted by a strong conical protuberance. I think, however, there can be little doubt that this conical protuberance is to be considered as an accidental outgrowth caused probably by the joint having been broken and afterwards healed. The first tarsal joint is very short, and the second is about as long as the second joint of the leg. At its extremity the last joint bears a comparatively feeble claw and two auxiliary claws. The joints of the legs have numerous but small and stout hairs ; they are at the swollen extremity of the second joint, and on the third and the fourth joints ; on the following joints they are much more numerous, but also a great deal more slender. On the two last joints, which have also stronger spines, for example on the side facing the claw, they are most numerous of all.

The female specimen shows very large genital pores at the considerably swollen distal extremity of the second joint of the leg. They are found ventrally on all the legs. The specimens seem to be very brittle, especially the younger ones, which had lost nearly all their legs.

Habitat.—This species was collected at three different stations not far from the coast of Patagonia.

Station 304. December 31, 1875. Lat. 46° 53' S., long. 75° 11' W. Depth of the sea, 45 fathoms. Bottom of the sea, sand.

Station 308. January 5, 1876. Lat. 50° 10' S., long. 74° 42' W. Depth of the sea, 175 fathoms. Bottom of the sea, mud.

Station 313. January 20, 1876. Lat. 52° 20' S., long. 68° 0 W. Depth of the sea, 55 fathoms. Temperature at the bottom, 8·8° C. Bottom of the sea, sand.

Observations.—In general, the shape of this *Pycnogonid* resembles that of *Phoxichilidium digitatum*, Böhm. However, in many respects, it may be easily distinguished from this and other species of *Phoxichilidium* hitherto described ; for instance, by the presence of ovigerous legs in the female, by the presence of auxiliary claws, by the number of joints (10) of the ovigerous legs, &c. Like most other species of the same genus, this species seems only to occur in shallow water (depth 45 to 175 fathoms) not far from the coast.

Phoxichilidium patagonicum, var. *elegans*, n. var..(Pl. XII. fig. 10).
Diagnosis.—Like *Phoxichilidium patagonicum*, Hoek, only much more slender.
Description.—

Length of the proboscis, .	. . .	.	3·5 mm.
Total length of the body, .	.	. . .	9·5 „
Length of the abdomen, .	.	. . .	2·5 „
Length of the leg of the third pair, .	.	. . .	28·5 „
Length of the ovigerous leg,	.	. . .	4 „

A young specimen has the different thoracic segments by no means so concentrated or robust as is the case in the specimens of *Phoxichilidium patagonicum*; the lateral processes are much more widely separated ; in general the length of the body, in comparison with that of the legs and of the proboscis, is much more considerable. The oculiferous tubercle is furnished with four eyes, but the two foremost are much larger than the two others. The length of their legs and their joints is not very different from that of *Phoxichilidium patagonicum*; the only difference being that the total length is comparatively less. The claws and the auxiliary claws are as in *Phoxichilidium patagonicum*. About the sex of this specimen I do not feel quite sure : most probably it is a young female.

Habitat.—Station 320. February 17, 1876. Lat. 37° 17′ S., long. 53° 52′ W. Depth of the sea, 600 fathoms. Bottom temperature, 2·7° C. Bottom of the sea, hard ground.

Observations.—The single specimen of this form resembles *Phoxichilidium patagonicum* so strongly that I hesitated long whether or not I should consider it as specifically distinct. My study of other species, younger and older specimens, has convinced me that, as a rule, as the animal advances in age, its slenderness increases. Now, in the present case, a young specimen shows considerable slenderness, while the full-grown female is much more concentrated. That it is a young specimen is proved by the rudimentary condition of the ovigerous leg. I therefore feel inclined to consider this form as a variety of my *Phoxichilidium patagonicum*. Considering the difference in depth of the stations at which that species and the present form were dredged, we have here most probably an instance of the influence of surrounding circumstances on the form of an animal.

Phoxichilidium mollissimum, n. sp. (Pl. XIII. figs. 6–9).

Diagnosis.—Body robust, lateral processes not very widely separated. Mandibles distinctly three-jointed, with curved, smooth, and not very long pincers. Ovigerous legs ten-jointed, without claw (probably present in both sexes). Auxiliary claws (?). Palpi represented by small rounded protuberances. Legs with silky hairs.

Description.—

Length of the proboscis,	9·5 mm.
Total length of the body,	28 ,,
Length of the abdomen,	9 ,,
Length of the ovigerous leg,	20 ,,
Length of the first six joints of the leg,	110 ,,

Of this interesting deep-sea Pycnogonid, unfortunately only a single specimen (much injured) was collected during the voyage of H.M.S. Challenger. There is not a single complete leg; and of the incomplete ones, with their three joints, there are in all only three present. However, this form is so highly characteristic that I think it possible to give, even from this defective specimen, a description which will be recognised by future investigators.

The body is robust; like the proboscis, the lateral processes, and the first joints of the legs, it is entirely smooth. The front of the cephalothoracic segment projects over the base of the proboscis. This front part bears dorsally the blunt oculiferous tubercle, with its rudimentary eyes, and quite anteriorly the large mandibles are inserted. These run parallel to one another, and are distinctly three-jointed. The basal joint is comparatively long, swollen at its base, and again at the extremity; the second is about two-thirds the length of the basal joint; the third joint is very short, and terminates in a pair of smooth, strongly-curved pincers, which, when closed, have a wide interval between them.

The proboscis is very stout, swollen a little in the middle, and also at the extremity; seen laterally, the swelling at the extremity appears rather stronger on the dorsal than on the ventral side. The triangular mouth is quite closed by the labial plates. Near the base of the proboscis the cephalothoracic segment bears on both sides a small blunt protuberance, which represents the palpus.

The ovigerous legs are inserted ventrally. Seen from that side, the cephalothoracic segment is distinctly divided into two segments, and the ovigerous legs originate from the first of these two segments. They are inserted on small processes, which represent the lateral processes of the ovigerous legs, and have ten joints. The first and third joints are short; the second is about twice as long; the fourth is comparatively stout and longer than the second; the fifth is as long as the second, and much more slender than the fourth; the sixth is shorter than the fifth, slender, but swollen at the extremity. The last four joints do not show any great differences in length; however, from the sixth to the tenth each joint is more slender than the preceding one. The hairiness of the different joints of the leg is not very great. Some hairs are found on the second to the sixth joints, the latter being, especially at the swollen extremity, furnished with some stronger hairs. The seventh joint is, near the extremity, armed with very long hairs; the eighth joint has them all over the surface; on the ninth joint the hairs are short and few; while the very slender tenth joint is furnished not only with some short hairs, but also with rows of stronger spines.

The lateral processes for the insertion of the legs are comparatively long. The abdomen is very long, being nearly cylindrical, and at the extremity a little swollen.

The second joint of the leg is twice as long as the first and the third. Of the three following joints, which are comparatively long, the first is a little longer than the second, and a little shorter than the third. With the exception of a row of not very numerous hairs placed at their extremities, the first three joints are smooth; the fourth shows already a small number of very thin hairs, which are much more numerous on the fifth, and extremely numerous on the sixth joints. On these last two joints, however, the hairs cover only half the surface longitudinally, whereas the other half has slender spines placed in distinct rows (fig. 9).

The single specimen of this species shows ventrally small genital pores on the second joint of the first, second, and third legs (the only legs present). From the smallness of these genital pores, the absence of ovaries in the fourth joint of the leg, and the shape of the ovigerous legs, I conclude that this specimen is a male.

Habitat.—The single specimen was brought up from

Station 237.   June 17, 1875.   Lat. 34° 37′ N., long. 140° 32′ E.   Depth of the sea, 1875 fathoms.   Bottom temperature, 1·7° C.   Bottom of the sea, mud.   .

Observations.—This species and the two following (*Phoxichilidium oscitans*, Hoek, and *Phoxichilidium pilosum*, Hoek) are nearly allied. They are furnished with ten-jointed ovigerous legs, present in both sexes, and three-jointed mandibles. The late Dr R. von Willemoes-Suhm erroneously mentioned these species as belonging to the genus *Zetes*, Kröyer. When comparing these species with true three-jointed mandibles, with *Phoxichilidium fluminense*, Kröyer, with two-jointed mandibles, and a distinct row of spines dorsally near the middle, and with *Phoxichilidium patagonicum*, Hoek, which has the basal joint divided into two when seen from the dorsal surface, and quite undivided when seen from the ventral surface, it becomes evident that this division, even when so distinctly developed as is the case with *Phoxichilidium mollissimum*, *Phoxichilidium oscitans*, and *Phoxichilidium pilosum*, does not justify us in considering these species as belonging to a different genus. Even should this be proposed, they could never be considered as species of *Zetes*; for that genus has ten-jointed palpi, whereas these organs in the present forms have become entirely rudimentary.

Phoxichilidium oscitans, n. sp (Pl. XIII. figs. 1–5).

Diagnosis.—Body robust, lateral processes not very widely separated. Mandibles three-jointed, with long and slender pincers. Ovigerous legs ten-jointed, without claws, probably present in both sexes. Auxiliary claws present. Palpi represented by rounded protuberances. Proboscis swollen in the middle, and considerably at the tip.

Description.—

Length of the proboscis, ...	8 mm.
Total length of the body, ...	21·5 ,,
Length of the abdomen, ...	6·5 ,,
Length of the ovigerous leg, ...	9·6 ,,
Length of the leg of the third pair, ...	97 ,,

A single specimen of this beautiful species was brought home by the Challenger. It is a species with a robust body, with long lateral processes, which are not widely separated, a very long abdomen, and a very stout proboscis. The first or cephalic part of the cephalothoracic segment is almost globular, and bears about its middle a blunt oculiferous tubercle with two rudimentary eyes, represented by brown spots, which are connected by a slender strip of pigment, the whole not unlike the form of what the French call a *pince-nez*. The cephalothoracic segment is nearly as long as the three other segments together. The abdomen is long, cylindrical, swollen at the extremity.

The proboscis is very stout; it is considerably swollen in the middle, and also at the extremity. This extremity is flattened at the front, and has a very large triangular mouth, the three lips of which are turned inwards. While the body is almost everywhere smooth, the front of the proboscis bears round the mouth not very long but comparatively strong hairs.

The mandibles are distinctly three-jointed. The first joint is the longest; the second is but little shorter; together they reach considerably beyond the end of the proboscis. The third joint is small, and bears a pair of extremely slender pincers, the movable one being much more strongly curved than the immovable one. At the end of the first joint a row of not very long but comparatively strong hairs is observed; those at the extremity and over the whole surface of the third joint are a little longer. The pincers are quite smooth.

The palpi are represented by very large globular protuberances, placed at both sides, near the base of the proboscis.

The ovigerous legs are inserted ventrally not far from each other. Seen from the ventral side, the cephalothoracic segment is much shorter than when observed dorsally; nor is there from that side any trace of a division into two joints to be seen, as is the case in *Phoxichilidium mollissimum*, Hoek.

The length of the joints of the ovigerous legs is exactly as in *Phoxichilidium mollissimum*; the first and third joints are short, the second is a little longer, the fourth longest of all and much thicker; the fifth joint is only a little shorter than the fourth, but considerably more slender; the sixth again is a little shorter than the fifth, and swollen at the distal extremity. The last four joints are short and nearly of equal length. The hairs on these last joints are not so long, but more regularly spread over the surface, than is the case in *Phoxichilidium mollissimum*. The tenth joint, however, shows nearly the same row of spines and the same short hairs as in the foregoing species.

The legs are very long. The relative length of the joints is the same as in *Phoxichilidium mollissimum*. The seventh joint is short; the eighth joint is a little curved. Both these joints are slender; together they are nearly equal in length to the second joint of the leg. The claw is very long, and furnished with two small auxiliary claws. That side of the eighth joint which faces the claw shows a row of stronger spines, the last of which is considerably longer than the others. Besides distinct rows of stouter hairs at the extremity of the joints, the number of hairs on the surface of the joints considerably increases from the fourth joint downwards, the greatest number being found on the distal part of the sixth joint.

The only specimen is probably a male. Its genital openings are small, and are placed ventrally on the second joint of every leg.

Habitat.—This beautiful deep-sea species was found at

Station 70. June 26, 1873. Lat. 38° 25′ N., long. 35° 80′ W. Depth of the sea, 1675 fathoms. Sea bottom, globigerina ooze.

Observations.—This species is nearly allied to the foregoing (*Phoxichilidium mollissimum*, Hoek), and also to the following species (*Phoxichilidium pilosum*, Hoek). It may, however, be easily distinguished from these species by the form of the proboscis and of the cephalothoracic segment, and by its extremely long legs.

Phoxichilidium pilosum, n. sp. (Pl. XIII. fig. 10–13).

Diagnosis.—Body not very robust, lateral processes widely separated. Mandibles distinctly three-jointed, with small straight pincers. Ovigerous legs ten-jointed, without claws, present in both sexes. Auxiliary claws present. Palpi represented by rounded protuberances. Proboscis cylindrical, tapering towards the extremity. Body (dorsally) and legs covered with extremely long and thin hairs.

Description.—

Length of the proboscis,	.	.	.	.	5 mm.
Length of the abdomen,	.	.	.	.	5 ,,
Total length of the body,	.	.	.	.	15 ,,
Length of the ovigerous leg,	.	.	.	.	8 ,,
Length of the leg of the third pair,	.	.	.	.	44 ,,

Of this species three specimens were collected during the cruise of H.M.S. Challenger;

there are two females and a male. The male is a great deal smaller than the larger of the two females. I therefore give the description from the larger female. The body is stout, but, as the lateral processes are widely separated, not very robust. The cephalothoracic segment is not very long. Quite in front of it the oculiferous tubercle is inserted so as to overhang the base of the mandibles. It is much elevated, conical, with two larger eyes directed forwards and two smaller ones directed backwards. The first and second true thoracic segments together are shorter than the cephalothoracic segment. The last thoracic segment is short, and bears at the extremity a very large abdomen, the length of which is nearly equal to that of the proboscis. The abdomen is a little swollen at the tip. Dorsally the surface of the body of the lateral processes and of the abdomen is furnished with very slender hairs, of which a distinct row is observed on the hinder margin of the different thoracic segments.

The two mandibles are three-jointed; the first two joints run parallel to each other. The first joint reaches as far as the end of the proboscis; the second joint is a little longer; the third is inserted on the second, with which it makes a right angle. The pincers of the two mandibles are directed horizontally towards each other; they are placed at a short distance from the end of the proboscis, are straight, and very short.

The proboscis is cylindrical, tapering towards the extremity, where the small mouth is situated. At the base of the proboscis the two protuberances representing the palpi are inserted.

The ovigerous legs are ten-jointed: the first and third joints are very small; the second is about twice as long; the fourth and fifth are the longest of all; the sixth joint is short; and of the four last joints the second is the longest and the last the shortest. They are covered with not very long but thin hairs, rows of stronger ones, as a rule, being found at the extremity of the joints.

The second joint of the legs is nearly twice as long as the first or third, and in the female considerably swollen at the extremity. The fourth joint is also considerably swollen in the female, and is longer than the first three joints together. The fifth joint is much more slender and also a little shorter than the fourth; the sixth is much longer and, at the same time, a great deal more slender. The two last joints together are not quite one-third the length of the sixth joint. The first tarsal joint is extremely short, the second about five times as long. From the first to the sixth all the joints are covered with very long and slender hairs, giving an extremely woolly appearance to the whole animal. The first tarsal joint is furnished with numerous stronger and smaller hairs; the second shows a row of distinct spines, the last of which is the largest, while the others diminish in size. The claw at the end of the leg is very large; it is accompanied by two extremely small auxiliary claws.

The females have very large genital openings ventrally on the swollen extremity of the second joint of all the legs. In the males I could observe the small genital pores only on the ventral surface of the two hindermost legs. They are here placed at the tip of a small tubercle, which is likewise absent on the two first pairs of legs. The genital pores of the males are almost quite covered by the surrounding hairs.

This hairy Pycnogonid was dredged by the Challenger at two different stations.

Station 147. December 30, 1873. Lat. 46° 16′ S., long. 48° 27′ E. Depth of the sea, 1600 fathoms. Temperature of the bottom, 0·8° C. Bottom of the sea, globigerina ooze.

Station 157. March 3, 1874. Lat. 53° 55′ S., long. 108° 35′ E. Depth of the sea, 1950 fathoms. Bottom of the sea, diatom ooze.

Observations.—The near relation which this beautiful deep-sea species bears to the two foregoing ones is evident. They form the true deep-sea representatives of the genus *Phoxichilidium*, Milne-Edwards, which probably will be found to have an extremely wide range.

Hannonia, n. gen.

Diagnosis.—Proboscis stout, inserted quite in front of the cephalothoracic segment. Mandibles rudimentary, small, two-jointed, chelate. Palpi wanting. Ovigerous legs, present in both sexes, ten-jointed.

Hannonia typica, n. sp. (Pl. XIV. fig. 8–11).

Diagnosis.—Body robust, proboscis long-ovate, truncated at the tip, and forming a narrow stalk posteriorly. Legs short, with a small first tarsal joint and a claw, without auxiliary claws.

Description.—

Length of the proboscis, . . .	4 mm.
Length of the trunk with the abdomen, .	7 „
Total length of the body, . . .	11 „
Length of the leg of the third pair, .	12 „

This curious sea-spider has a robust body; the cephalothoracic segment is not very large, and, like the two following segments, its hinder margin is furnished with an elevated ridge; on these ridges a row of small hairs is inserted. The oculiferous tubercle is blunt, and placed in the middle of the cephalothoracic segment. It is furnished with four comparatively large black-coloured eyes. The dorsal surface of the lateral processes, like that of the last thoracic segment immediately in front of the insertion of the abdomen, shows rounded protuberances, on the surface of which small hairs—like those of the ridges on the hinder side of the thoracic segments—are placed. The abdomen is not very long, but stout and swollen at its extremity. Its surface is likewise covered with numerous small hairs. The form of the proboscis is ovate; at the anterior end it is flattened and truncated. The mouth is found in the middle of this truncated surface; it is triangular, with comparatively large lips. At the back the proboscis is narrower, the narrow part thus forming a sort of stalk-like process on which the ovate front part is borne. The surface of the proboscis, when seen with the naked eye or slightly magnified, is entirely smooth.

The mandibles are small and rudimentary; they consist of a short basal joint and a quite rudimentary second joint, armed with rudimentary pincers. The length of the mandibles is about one-fourth the length of the proboscis.

The ovigerous legs are not very strong ; they are ten-jointed. The first three joints are short; the fourth and the fifth joints are the longest; the sixth is about two-thirds the length of the fifth; and of the last four joints the first is by far the longest, and the third the shortest. A small claw is found at the extremity of the tenth joint. All the joints are furnished with small hairs; the last four joints are not armed with denticulate spines, but with not very strong straight spines, scattered rather irregularly over the whole surface of the joints.

The legs are short. The first three joints are extremely short; the three following are longer and nearly of the same length, which is shorter than that of the first three joints together. The first tarsal joint is very short; the second comparatively long and feebly curved, it bears at the extremity a small strongly-curved claw, which is sickle-shaped, and not accompanied by auxiliary claws. All the joints of the legs are furnished with numerous minute spines, placed in regular rows ; the fifth and sixth joints, however, are also armed with a distinct row of tubercles, each bearing a small but strong spine at the tip (fig. 11).

The only specimen of this species is a female. It has very large genital pores on the second joints of all the legs. The ovaries are found reaching as far as the sixth joint of the leg. The ovarian eggs are exceedingly numerous, but comparatively large.

This interesting Pycnogonid was found on the shore at Seapoint, near Capetown.

Observations.—It is a true shore inhabitant, and forms among the species without palpi the transition from those with (*Pallene* and *Phoxichilidium*) to those without mandibles (*Pycnogonum* and *Phoxichilus*). To the first of the latter genera · (*Pycnogonum*) it is, I believe, very nearly allied—viz., by the robustness of the body and by the presence of the protuberances (which I showed in my paper published in 1877, Ueber Pycnogoniden, to be outgrowths of the skin, richly armed with tactile organs) on the dorsal surface of the body and of the lateral processes. The want of auxiliary claws in both genera is also striking. Distinct differences are furnished by the presence of mandibles, and of ovigerous legs in the female of my *Hannonia typica;* I have already pointed out above, however, that I do not consider these differences very important.

APPENDIX I.

DESCRIPTION OF THE SPECIES DREDGED DURING THE CRUISE OF THE "KNIGHT ERRANT."

During the recent cruise of the "Knight Errant," organised by Prof. Sir Wyville Thomson to acquire a more accurate knowledge of the abnormal distribution of temperature in the Faroe Channel, numerous Pycnogonids were collected by trawling. As this cruise bears upon the voyage of the Challenger and the study of the results of her voyage, I was asked to give also a description of these forms for this report.

Nymphon strömii, Kröyer.

> *Nymphon strömii*, Kröyer, Bidrag til Kundskab, &c., Natürh. Tidskr., N. R., i. 111, 1845.
> *Nymphon gracilipes*, Heller, Crustaceen, Pycnogoniden, und Tunicaten der K. K. Oester. Ungar. Nordpol. Exped. Denkschr. der Math., Naturw. Classe der Kaiserlichen Akad. der Wiss., xxxv. 40, 1875.
> *Nymphon strömii*, Kröyer, Miers, Ann. and Magazine, 4th series, vol. xx. p. 109, 1877.
> *Nymphon gracilipes*, Heller, G. O. Sars, Prodromus descriptionis Crustaceorum et Pycnogonidarum, quae in expeditione Norvegica, anno 1876, observavit, Arch. f. Math. og Naturvid., ii. 265, 1877.
> *Nymphon strömii*, Kröyer, Wilson, Pycnogonida of New England, Transact. Connect. Acad., vol. v. p. 17, pl. vi. fig. 1a–1h, 1880.

This beautiful and distinct species is accurately described by Kröyer, and also by Wilson. Its synonymy and wide geographical range I have discussed at some length in the description of the Pycnogonids collected during the two cruises of the Dutch schooner "Willem Barents" in the Barents Sea, which at this moment is in the hands of the printer, and will be published probably before the end of the year (1880).[1]

The dimensions of the "Knight Errant" specimens are much smaller than those of the specimens described by Heller and myself, and even smaller than those which Wilson has got from the neighbourhood of the North American coast.

The extent[2] of the largest "Knight Errant" specimen is not quite 100 mm. The

---

[1] Supplement-Band of the Niederländisches Archiv für Zoologie, Leiden, E. J. Brill.
[2] "Extent is the distance from tip to tip of the outstretched legs" (Wilson, *loc. cit.*, p. 5).

depth from which this species was brought up is from 515 to 540 fathoms. The stations where it was found are the following :—

Station No. 5 (cruise of the "Knight Errant"). Lat. 59° 26′ N., long. 7° 19′ W. August 11, 1880. 515 fathoms. Warm area. Two specimens.

Station No. 7 (cruise of the "Knight Errant"). Lat. 59° 36′ N., long. 7° 18′ W. August 12, 1880. 530 fathoms. Warm area. Two specimens.

Station No. 8 (cruise of the "Knight Errant"). Lat. 60° 3′ N., long. 5° 51′ W. August 17, 1880. 540 fathoms. Cold area. Ten specimens.

In this animal, therefore, we have an example of one inhabiting the cold and warm areas on both sides of the ridge rising in the Faroe Channel to within 300 fathoms of the surface.[1] This agrees very well with the facts of the geographical distribution of our species; it is a common inhabitant of the depths of the Arctic Sea, but it is also by no means rare in the deeper water of southern latitudes, especially in the neighbourhood of the American coast.

Nymphon grossipes, Oth. Fabricius.

Nymphon grossipes, O. Fabricius, Fauna Groenlandica, p. 229, 1780.   (See p. 44 of this report.)

A single specimen of this species was dredged at

Station 8 (cruise of the "Knight Errant"). Lat. 60° 3′ N., long. 5° 51′ W. August 17, 1880. 540 fathoms. Cold area.

For the geographical distribution of this species I refer to the list of species at p. 20.

Nymphon macronyx, G. O. Sars (Pl. XV. figs. 1–7).

Nymphon macronyx, G. O. Sars. Prodromus descriptionis, &c., Archiv. f. Math. og Naturvid., ii. 265, 1877.

Of this interesting inhabitant of the cold area of the Faroe Channel about thirty specimens were dredged during the cruise of the "Knight Errant." As hitherto neither a full description nor any figure of this species has been published, I wish to give both here.

Only a short diagnosis of this species has been published by Professor G. O. Sars. From this, and from the pencil drawing he had the kindness to send me, the species is easily recognised. However, in some respects, I observed slight differences from the diagnosis of Professor G. O. Sars. Probably these will be found to arise from the fact that the species had been submitted only to a preliminary investigation by the celebrated Norwegian zoologist.

Nymphon macronyx, G. O. Sars, is a somewhat robust *Nymphon*, having the second joint of the palpi longer than the third, the first tarsal joint not quite half the length of the second tarsal joint, and having a very long claw at the end of every leg and

[1] Nature, September 2, 1880, C. Wyville Thomson, the Cruise of the "Knight Errant."

extremely small auxiliary claws. Moreover, it is characterised by a curiously-shaped oculiferous tubercle. Its dimensions are as follows :—

Length of the proboscis, .	. . . .	2·3 mm.
Total length of the body, ♀,	. . . . .	5·4 „
Total length of the body, ♂,	. . . . .	5·7 „
Length of the ovigerous leg, ♀, .	. . . .	6·4 „
Length of the ovigerous leg ,♂, .	. . . .	7·1 „
Length of the leg of the third pair,	. . . .	18 „

The body is almost quite smooth, while the appendages are richly furnished with hairs. The slenderness of the body is not very great; the lateral processes, however, are widely separated (Pl. XV. fig. 1). The cephalothoracic segment is shorter than the proboscis; it is narrow in the middle, while it shows a considerable swelling at the beginning and at the end. The oculiferous tubercle is situated above the insertion of the ovigerous legs. Seen laterally this shows the conical and pointed shape described by Professor G. O. Sars; but seen from the front it is considerably flattened and broad, terminating in two divergent points. It is furnished with four distinct eyes (Pl. XV. fig. 2). The proboscis is almost quite cylindrical. The abdomen is short.

Of the appendages I have figured the mandibles in fig. 3 on Plate XV. The second joint is nearly triangular, and almost its whole surface is covered by strong hairs. At the front, one of the angles of the triangle terminates in the immovable claw, which is a great deal shorter than the movable one. Both claws bear a row of spines and are strongly curved at their extremities.

The drawing I have given in fig. 4 of the palpi renders, I believe, a description needless.

The ovigerous leg of the female is shorter and feebler than that of the male; moreover, it has the fourth and fifth joints quite straight, whereas the same joints in the male describe distinct curves. The relative length of the joints is the same as in most other species of the genus *Nymphon*. In the male the sixth joint is furnished with a small pointed tubercle, which I did not observe in the female. The males bear the eggs on the fourth and fifth joints; they are large and collected in one or two packets on both legs. The four last joints show a row of denticulated spines of the shape figured on Plate XV. fig. 5. The claw at the end of the ovigerous leg is strong and pectinated at the extremity.

The second joint of the legs is twice as long as the first and the third, the fourth joint is as long as the fifth, the sixth is only a little longer. The first tarsal joint is not quite half as long as the second. The claw is very long, almost as long as the second tarsal joint. Very small auxiliary claws, easily overlooked, are situated on both sides of the great claw (fig. 7). The second joint of the female is considerably swollen at the end, where it shows on every leg a large genital pore of an oval shape; the fourth joint in the same sex is also swollen. The hairs on the first four joints in both sexes are not very numerous, whereas, beginning with the fifth, the last joints are richly furnished with hairs. The side of the eighth joint facing the claw shows a row of regular strong spines. The genital pores of the males are smaller than those of the females; they are present only on the two posterior pairs of legs.

As for the geographical distribution of this species, G. O. Sars dredged it in lat. 62° 44′ 5″ N., long. 1° 48′ E., in comparatively deep water (412 fathoms), in the cold area. He found there only four specimens, whereas a single haul with the trawl in the Faroe Channel yielded among a thousand specimens of *Nymphon robustum*, Bell, upwards of thirty specimens of the species in question. This occurred at

Station No. 8 (cruise of the "Knight Errant"). Lat. 60° 3′ N., long. 5° 51′ W. August 17, 1880. 540 fathoms. Cold area.

This station is not very far from the place where it was dredged by Professor G. O. Sars, and as this is the only instance, so far as I know, of this species having been collected previous to the cruise of the "Knight Errant," most probably it has a very restricted distribution.

Nymphon robustum, Bell.

> *Nymphon robustum*, Bell, Belcher's Last of the Arctic Voyages, vol. ii. p. 409, 1855, Tab. xxxv. fig. 4.
> *Nymphon abyssorum*, Norman, Wyville Thomson, Depths of the Sea, p. 129, 1873.
> *Nymphon hians*, Heller, Crustaceen, Pycnogoniden und Tunicaten der K. K. Oester. Ungar. Nordpol. Exped. Denkschriften der Wiener Akademie der Wiss., xxxv. 41, 1875.
> *Nymphon robustum*, Bell, G. O. Sars, Prodromus Crustaceorum et Pycnogonidarum, Arch. für Math., og Naturvid., ii. 265, 1877.

An ample discussion of the synonymy of this species, and a description of those parts of the body which have hitherto been overlooked, I have given in my paper on the Pycnogonids of the cruises of the "Willem Barents," to which I have referred above. An immense quantity of this true cold area species was dredged during the recent cruise of the "Knight Errant." Mr Murray writes to me that this was the greatest haul of Pycnogonids he ever observed. It is a blind species, and along with it were trawled a considerable number of specimens of *Nymphon macronyx* which have distinct eyes, about ten specimens of *Nymphon strömii* also furnished with eyes, one specimen of *Nymphon grossipes* with eyes, and four specimens of *Colossendeis proboscidea*, Sabine (sp.), which is again without eyes. The number of specimens with eggs is not very considerable, and there is not one which shows the numerous highly developed young ones clinging to the ventral side of the body of their parent as is the case with some specimens from Barents Sea.

Finally, I wish to point out that the dimensions of the "Knight Errant" specimens are considerably smaller than those of specimens from higher latitudes. As I have mentioned above, this is also the case with the specimens of *Nymphon strömii*.

A species of *Scalpellum*, which, so far as I know, has not been observed hitherto, is a common commensal on the legs of *Nymphon robustum*, Bell.

This species was dredged at

Station 8 (cruise of the "Knight Errant"). Lat. 60° 3′ N., long. 5° 51′ W. August 17, 1880. Depth of the sea, 540 fathoms. Cold area. Perhaps 1000 specimens.

Station 2 (cruise of the "Knight Errant"). Lat. 60° 29′ N., long., 8° 19′ W. July 28, 1880. Depth of the sea, 375 fathoms. Warm area. One specimen only.

This single specimen from the warm area must, most probably, be considered as one which has strayed from the cold area. As far as I know, the species has not been observed at a lower latitude than 60° N.

Nymphon robustum, Bell.

Colossendeis proboscidea, Sabine (sp.).

<blockquote>
Phoxichilus proboscideus, Sabine, Marine Invertebrate Animals in a Supplement to the Appendix of Captain Parry's Voyage for the Discovery of a North-West Passage in the years 1819–1820, London, John Murray, 1824, p. ccxxvi.

Colossendeis borealis, Jarzynsky, Præmissus Catalogus Pycnogonidarum inventarum in mari glaciali ad oras Lapponiæ rossicæ et in mari albo, anno 1869 et 1870. Annales de la Soc. des Naturalistes de St Pétersbourg, vol. i., 1870.

Colossendeis proboscidea, Sabine (sp.), G. O. Sars, Prodromus descriptionis, &c., Archiv. für Math. og Naturvid., ii. 268, 1877.
</blockquote>

Of this interesting species, the first of the genus *Colossendeis* that was observed, four specimens were trawled at Station 8, together with many specimens of *Nymphon robustum*, *Nymphon macronyx*, &c. (See above.) There is one very young specimen, and the three others are females. For figures and a full description of this species I again refer to my paper on the Pycnogonids of Barents Sea.[1]

The specimens trawled by the "Knight Errant" are not quite so large as those from Barents Sea.

Station 8 (cruise of the "Knight Errant"). Lat. 60° 3′ N., long. 5° 51′ W. August 17, 1880. Depth of the sea, 540 fathoms. Cold area.

Pycnogonum litorale, Ström (sp.).

> *Phalangium litorale*, Ström, Physisk og oeconomisk beskrivelse over fogderiet Söndmör, beligende in Bergens Stift i Norge, 4°. Soröe, 1762-66, pl. i. fig. 17.
> *Pycnogonum litorale*, O. Fabr., Fauna Groenlandica, p. 223, 1780.
> *Pycnogonum litorale*, Müller, Zoologia Danica, iii. 68, pl. cxix. figs. 10-12, 1789.
> *Pycnogonum litorale*, Kröyer, Nat. Tidsk. Ny Række, i. p. 126, 1845.

Of this very common species one specimen was dredged at 53 fathoms. It occurs only in the neighbourhood of the coast, and ranges to the north as far as the White Sea, where Jarzynsky (Præmissus Catalogus, &c., *loc. cit.*) collected it on the coast of Russian Lapland, and as far south as the coast of the Mediterranean. Westward it is common at different places on the North-American coast, and it also abounds on the east coast of the Atlantic—as on the English, Dutch, French coasts, &c. Slater (Ann. and Mag. of Nat. Hist., v. series, vol. iii., 1879) describes a variety of this species—it is a little more slender—collected on the coast of Japan. Most probably, therefore, the species will also be found to occur along the whole northern coast of Siberia.

The single specimen trawled in the neighbourhood of the Scottish coast is a male with distinct ovigerous legs. It was dredged at

Station No. 3 (cruise of the "Knight Errant"). Lat. 59° 12′ N., long., 5° 51′ W. August 3, 1880. Depth of the sea, 53 fathoms.

---

[1] This same species has been recently described by Mr E. J. Miers under the name *Anomorhynchus smithii*, n. gen., n. sp., from specimens collected by Mr Leigh Smith a little to the south of Franz-Josef Land (Annals and Magazine of Natural History for January 1881, p. 50, pl. vii. figs. 6-8). (Note inserted during the correction of the last proof.)

APPENDIX II.

CONTRIBUTIONS TO THE ANATOMY AND EMBRYOLOGY OF THE PYCNOGONIDA.

Our knowledge of the anatomy and embryology of the sea-spiders is very insufficient ; of those living in shallow water we know but little, and of the deep-sea forms nothing. Whereas *a priori* it is evident, that in general the deep-sea animals will exhibit the same anatomical structure, and pass through the same development as the littoral or shallow-water forms ; it is also clear, on the other hand, that a comparison of the anatomy of animals inhabiting very different depths might lead, at least in the case of some organs, to very interesting results.

As the rich material collected during the voyage of H.M.S. Challenger enabled me to study the anatomy of at least some genera (*Nymphon* and *Colossendeis*), I eagerly made use of this opportunity ; in the first place, in the hope of increasing our knowledge of the morphological structure of the group, so that the question of their position in the zoological system might perhaps be settled ; and in the second place, to try in this way to illustrate the mode of life of those deep-sea animals which belong to our group.

My original intention of going through the whole anatomy of the Pycnogonids I have given up, seeing that, however good the condition of the material might be, yet in regard to some organs,—intestine, heart, &c.,—and for the histological structure of most organs, it by no means takes the place of fresh material. Moreover, as I learned after a great part of my anatomical researches was finished that Dr Dohrn's Monograph is forthcoming, I determined to limit the publication of my researches to those organs which had suffered least from having been in alcohol for so many years. These are the integument, with its glands ; the nervous system, with the sensory organs ; and the reproductive organs. In regard to the intestine, the heart, &c., only some incidental observations were made, which, in so far as they are thought important enough, will be recorded also.

1. *Integument.*—The integument of the Pycnogonids is only known from the publications of Zenker (1855),[1] myself (1877),[2] and Dohrn (1879).[3] Zenker was the first to observe the numerous cavities in the chitinous cuticle of *Pycnogonum litorale*. I, however, had the good fortune to demonstrate that these cavities communicated by

[1] Zenker.—Untersuchungen über die Pycnogoniden, Müller's Archiv, 1852.
[2] Hoek.—Ueber Pycnogoniden, Niederländisches Archiv für Zoologie, iii., 1877.
[3] Dohrn.—Neue Untersuchungen über Pycnogoniden, Mittb. Zool. Stat. Neapel., i., 1879.

means of very narrow canals with the exterior, and that they occur in all the genera of Pycnogonids (at least in those I had then studied—*Nymphon, Pallene, Phoxichilidium,* and *Pycnogonum*). Moreover, I pointed out that as no respiratory organs are present in the Pycnogonids, respiration must necessarily be integumentary; it was my conviction in 1877, as it is still, after a minute investigation of the Challenger material, that the principal function of these canals is to serve for respiration. Contrary to this opinion, Dohrn asserts that the cavities, with the pore-canals, which he says, were rightly described by me, "zur Aufnahme von Hautdrüsen dienen." To settle this question I investigated the structure of the integument of many species belonging to different genera. I studied it in *Nymphon hamatum, N. longicoxa, N. brachyrhynchus,* and *N. brevicaudatum;* in *Colossendeis leptorhynchus, C. gigas,* and *C. proboscidea;* in *Ascorhynchus glaber* and *A. orthorhynchus;* in *Pallene australiensis;* and in *Phoxichilidium patagonicum, P. pilosum,* and *P. insigne.*

Notwithstanding that my researches were especially directed to this point, I only once succeeded in observing the glands referred to by Dohrn, and although I grant it is possible that in some cases this may be owing to the condition of the animals, yet I feel sure that as a rule these glands are not present.

A short description of the integument may find a place here. It always consists of a subcuticular layer (epithelium), and of the chitinous cuticle. The subcuticular epithelium is of a protoplasmic nature, with nuclei imbedded in it[1] (Pl. XVI. figs. 1 and 17, *f*); the chitinous cuticle in the different species shows a very different thickness, and always presents a stratified appearance. It is never calcified, and, as a rule, is coloured yellow by picrocarmine. Often, however, it shows two distinct laminæ; an internal very thick one, coloured violet by the picrocarmine, and consisting of numerous alternately lighter and darker strata, and a comparatively thin external one, which assumes a yellow colour when treated with picrocarmine (Pl. XVI. fig. 1). To strengthen the often extremely long and slender joints of the legs (especially the thighs and the two tibial joints), the chitinous cuticle is often furnished internally with one (*Ascorhynchus glaber,* Pl. XVI. fig. 9) or two (*Nymphon hamatum, Phoxichilidium insigne*) longitudinal ridges, which project into the interior of the leg. The form of these ridges on a transverse section is by no means always the same, as may be seen from the figures 6, 9, 11, 16, 17, and 18 on Pl. XVI. The septa of connective tissue, which in most genera divide the cavity of the joints of the leg (*Colossendeis, e.g.,* fig. 16, Pl. XVI.), often have a point of attachment in these ridges.

As a rule the chitinous cuticle of the Pycnogonids is perforated by two kinds of cavities, the one of an irregular conical shape, terminating externally in a narrow pore-canal; the other much narrower, and rather more cylindrically shaped, is filled

---

[1] The subcuticular epithelium of *Pallene australiensis,* Hoek, is richly furnished with a dark brown pigment. I did not observe this in any of the other species.

with a protoplasmic substance, often containing nuclei. No doubt it is the protoplasmic epithelium which fills these latter canals. A nerve passes through them, and terminates at the surface of the cuticle in what Dohrn (*loc. cit.*, p. 38) calls a " Borstenapparat." Dohrn never observed a single seta[1] at the end of these canals ; but always two or more (sometimes even a rosette of eight or nine) together. This observation of Dohrn's seems to be most accurate ; as a rule I found the integument of the species of *Nymphon* furnished with forked setæ (Pl. XVI. figs 1, 2, 4, 7), one of the setæ being often split again (fig. 3) ; that of *Pallene australiensis* shows also forked setæ ; the integument of *Ascorhynchus glaber* is also furnished with double setæ, which are here extremely small and rudimentary. One of the species of *Phoxichilidium* (*P. patagonicum*) shows a combination of four or five (Pl. XVI. fig. 17), while in another species (*Phoxichilidium pilosum*) two combined setæ are always observed.

In *Colossendeis* setæ are totally wanting ; and thus in this respect the genus *Ascorhynchus*, with its rudimentary setæ, stands between *Nymphon* and *Colossendeis*.

According to Dohrn,[2] this pore-canal, which terminates in these forked setæ, " often " takes its origin from one of the integumentary cavities, which he saw filled up with glands. Dohrn therefore considers these setæ as tactile organs, having probably the special function of causing on irritation, by reflex action, the secretion of a poisonous fluid by the glands, which are situated in the conical cavities of the integument. I think this a very ingenious supposition, but I wish to state in opposition to it, that according to what I have seen of the matter—(1) as a rule the cylindrical and narrow pore-canals do not originate in the conical cavities ; and (2) that the occurrence of the glands in the conical cavities is the exception, and that, as a rule, these cavities are empty or partly filled up with a protoplasmic substance, nuclei, blood-corpuscles, &c. In the different species of *Nymphon* it hardly ever happens that the pore-canal with the setæ takes its origin in a conical cavity ; in the species of *Phoxichilidium* it does not seem to be so rare (Pl. XVI. fig. 17), even in these species, however, it is by no means the rule.

With regard to the form of the conical cavities, in the first place it must be mentioned, that they have a most regular conical shape in the different species of the genus *Nymphon*, and also in some species of *Pallene*. In *Phoxichilidium* they are of a more elongated form, and often a small lateral branch passes from the main canal near the extremity (Pl. XVI. fig. 17, *b*). The genus *Colossendeis* shows these cavities of a much more irregular shape. As a rule every cavity is bifid, and terminates in two narrow pore-canals (Pl. XVI. fig. 12, *c*). The cavities are usually almost quite filled up with protoplasmic substance. In this genus I once observed distinct cells, with large

---

[1] Huxley calls " setæ " all the hair-like processes from the fine microscopic down to stout spines, which are found on the outer surface of the cuticle (Crayfish, London, 1880, p. 197). I use the word here, and on the following pages, in a much more restricted sense ; having already used the words hairs and spines for the integumentary appendages, I call " setæ " those which I consider as being more particularly of a sensory nature.

[2] *Loc. cit.*, p. 38.

nuclei, in these cavities. This was in the first tibial joint of the leg of *Colossendeis leptorhynchus;* however, I think there can be no doubt that these same cells will also be found in the other joints of the legs of this species. Each cell terminates in a long and slender appendage, which probably extends to within a small space of the opening of the canal. Most probably these are the glands which, according to Dohrn, are always present in these integumentary cavities. I have figured these glands in fig. 1 of Plate XVIII.; *i* is a part of the wall of the intestinal crœcum, which runs through the joint; *c, c* are parts of the septa of connective tissue, which seem to form here separate chambers in connection with the different cavities. These chambers contain numerous blood-corpuscles of an irregular spool form, and towards the pore-canal are furnished with two (*d*), in an other cavity three (*d'*) glandular cells, with very large nuclei. The specimen, the integument of which shows these glands, is a female.

Returning to these integumentary cavities, and their ordinary, viz., their respiratory, function, I have still to mention that I found them in many species with many blood-corpuscles in their interior, and that often also a nerve is seen which sends a very thin branch into them. These I observed more accurately in *Colossendeis proboscidea,* Sab. (See later, under peripheral part of the nervous system.)

The number of these cavities is different in the different species. I counted them in transverse sections of the fourth joint of the leg in some fourteen species, belonging to five genera, and compared them with the girth of the joint. This I did to ascertain if there was any relation between the number of these cavities and the depth at which the species lives. That such a relation does not exist, and that the greater or smaller number of these cavities is one of the properties of the natural groups (genera) of the Pycnogonids, is shown, I believe, by the following table :—

Name of the Species.	Circumference in millimeters of the fourth joint of the leg.	Number of integumental cavities in a transverse section.	Number per millimeter.	Depth in fathoms.
Nymphon brevicaudatum, Miers, . . .	3·25	5	1·54	73
Nymphon brachyrhynchus, Hoek, . . .	1·52	22	14·4	83
Nymphon robustum, Bell, . . .	4·6	27	5·9	458
Nymphon longicoxa, Hoek, . . .	2·35	12	5·1	1100
Nymphon hamatum, Hoek, . . .	3·47	37	10·7	1488
Ascorhynchus orthorhynchus, Hoek,	3·4	90	26·5	130
Ascorhynchus glaber, Hoek, . . .	3	56	18·6	1375
Colossendeis proboscidea, Sab. (sp.),	10·2	106	10	540
Colossendeis leptorhynchus, Hoek, .	3·37	63	18·7	1126
Colossendeis brevipes, Hoek, . . .	3·15	62	20	2650
Pallene australiensis, Hoek, . . .	2·4	20	8·3	79
Phoxichilidium insigne, Hoek, . . .	1·57	41	26	14
Phoxichilidium patagonicum, Hoek, .	5·2	112	21·5	117
Phoxichilidium pilosum, Hoek, .	4·17	45	10·8	1790

When comparing the number of these cavities in the fourth joint of the leg with that of the same organs in a transverse section of the body,—for example, between two lateral processes, where the circumference of the body is in some species nearly the same as that of the fourth joint of the leg,—I observed almost the same number of cavities. This was the case at least in *Nymphon hamatum* and in *N. brachyrhynchus;* whereas in the other species more or less considerable differences were observed, the number of these cavities in some species being greater in the legs; in others, on the contrary, round the body.

With regard to the hairs and spines on the surface of the body, I have already pointed out above that, as a rule, species occurring at great depths are rather smooth, whereas those from shallow water are furnished with numerous hairs and spines. Thus *Nymphon longicoxa* and *N. hamatum* have the surface almost quite destitute of spines; *Nymphon brevicaudatum* and *N. hirtipes* occurring at moderate depths, the former not exceeding 120 fathoms, the latter never reaching 300 fathoms, and generally found in considerably shallower water are the most hairy species of the genus. On the contrary, *Nymphon brachyrhynchus*, occurring at depths not exceeding 120 fathoms, is almost as smooth as *Nymphon hamatum*.

The species of *Colossendeis*, and especially the three more accurately studied by me, show almost a perfectly smooth surface. The sensory setæ are wanting also in these species; and the few spines which are present are very short and conical (Pl. XVI. fig. 13). Of these three species, two are true deep-sea inhabitants; but the third (*Colossendeis proboscidea*, Sab. (sp.), as a rule, is found at a depth not exceeding 200 fathoms. Of the species of *Ascorhynchus*, the smooth *A. glaber* is found at a depth of 1375 fathoms; but the surface of *Ascorhynchus orthorhynchus* is also not very hairy, yet this species occurs at a depth of only 130 fathoms.

The shallow water genera *Achelia* and *Ammothea* are extremely hairy, whereas in the genus *Phoxichilidium* some of the deep-sea species (*P. pilosum* and *P. mollissimum*) show a particularly hairy surface. Both the spines and the setæ are in these species of a very remarkable length. Finally, *Pallene australiensis*, occurring at a depth of 38 to 120 fathoms, shows again the smooth surface of a true deep-sea species.

The form of these spines is also very different, but I think it is not necessary to describe them. In some species the spines are not smooth, but serrated; as, *e.g.*, in the case of *Nymphon brevicaudatum*, Miers, and *Pallene australiensis*, Hoek (Pl. XI. figs. 6, 7); and as spines having a very curious shape I have pointed out already those of the sixth joint of the ovigerous leg of *Nymphon longicoxa*. No doubt, these must be of great use to the animal in holding the egg-masses, and perhaps also in furnishing a good point for the young ones to cling to. Particularly interesting are also the so-called denticulate spines in the four last joints of the ovigerous legs of most species. I may refer to the descriptive part of this report for an account of their extremely different forms, their numbers, and their arrangement.

Different as the forms of these spines may be, their minute structure is always the same, and, I think, quite identical with that of any other spine. The cuticle is perforated at the place where the spine is inserted, and a thin and flexible part of this cuticle keeps the spine in its place; a socket is thus formed in which the spine easily moves. The spine itself is, near its insertion, cylindrical and hollow, and its cavity is entirely or partly filled with a protoplasmic substance, which is in continuity with the epithelium of the integument. Towards the extremity the spine is flattened, chitinous, and no longer hollow; the exterior margin of this flattened part is serrated (*Nymphon*), or provided with extremely small teeth (*Colossendeis*). Originally I considered these spines as being of a sensory nature; but afterwards, as I was convinced of their chitinous composition, I changed my opinion. However, as it is not difficult to trace a nerve, at least in some of the more transparent species, penetrating these spines, they may still be considered as, to a certain extent, organs of feeling. On the other hand there are, perhaps, far more important functions to be fulfilled by the ovigerous legs with the aid of these denticulate spines, viz., those of seizing the food, and, last, not least, of holding the animal of the other sex during the act of copulation. In most species where denticulate spines occur the four last joints of the ovigerous legs often lie rolled up spirally, with the rows of denticulate spines turned inwards. These joints if wound round one of the legs or any other part of the body of the animal with which it copulates, would necessarily secure a very strong adhesion in consequence of the rows of spines.

In close relation with the integument are the glands, which occur in different appendages of the body.

1. The glands of the palpi. These I observed in *Discoarachne brevipes*, where they occur in the third joint of the palpi, and probably their secretion is poured out through a sieve-like perforated spot at the end of the second joint (Pl. VII. fig. 10). In the palpus of *Ascorhynchus* such a gland is also present. It is situated in the fifth joint, and has the form of a long sack, whose wall is lined by small and very numerous glandular cells. It is attached to the wall of the joint of the palp by means of numerous threads of connective tissue, and it opens towards its distal extremity by means of a rounded pore. In *Ascorhynchus orthorhynchus* this pore is placed at the tip of a small conical excrescence; but in *Ascorhynchus glaber* I did not observe this knob. Probably the same glands occur also in other genera—*e.g.*, in *Colossendeis*—but I could not ascertain their presence.[1]

2. The glands of the ovigerous legs. These I observed in *Nymphon*, but their minute structure can only be studied in fresh specimens. They open into a small pore, not far from the beginning of the fourth joint of the leg. Each gland seems to consist of the

---

[1] The glands of the palpi are mentioned in Dohrn's paper of 1879. He says of the nerve of the palpus: "Er umfasst auf seinem Laufe ein sonderbares, bisher unbekannt gebliebenes Excretionsorgan" (*loc. cit.*, p. 31).

true glandular part and of a wider part, which acts probably as a receptaculum and opens into the fine pore.[1]

3. The glands in the fourth joint of the legs of the males. These I observed in the following species:—*Nymphon hamatum, Nymphon brachyrhynchus, Ascorhynchus glaber, Colossendeis leptorhynchus, Colossendeis proboscidea, Colossendeis megalonyx, Oorhynchus aucklandiæ, Phoxichilidium insigne,* and *Pallene australiensis.* Whether or not they occur in the thighs of the males of all the species I dare not assert. I can only say that I did not find them in the thighs of the following species:—*Nymphon robustum, N. brevicaudatum, N. longicoxa, N. grossipes,* and *Colossendeis gracilis.* This may, however, be the consequence of these glandular masses being present only during a short period of the year—for example, only in the breeding season.

These are the glands which were observed by Dohrn. According to him (*loc. cit.,* p. 36) they occur only in the male sex, and are found in the fourth joint of the legs, viz., in those joints in which in the females the ovaries are most strongly developed. Dohrn, moreover, tells us that these glands are extremely variable in their appearance; while in *Ammothea* they are furnished with a single duct only, in *Phoxichilus* there are fifteen smaller openings. Dohrn supposes that the function of these glands is to secrete a viscous fluid, wherewith the males agglutinate the eggs which have been laid by the females, and attach them to their ovigerous legs.

As to the function of these glands, I was unable to make any observations from the material preserved in spirits and brought home by H.M.S. Challenger. With regard to their structure the following may be stated:—The glands are always composed of a skeleton of connective tissue, the meshes of which are or are not filled with nucleated cells, which doubtless are the true glandular cells. When these glands have the meshes filled up with the nucleated cells (Pl. XVI. figs. 5 and 10), the skeleton of connective tissue is not easily discerned; while in those cases in which the meshes are empty (Pl. XVI. fig. 15) the structure of the connective tissue is easily observed. In the form of the glandular cells small differences were also observed. In *Ascorhynchus glaber,* they are fusiform and pointed at both extremities; in *Nymphon hamatum,* they are rounded, but in both cases distinct nuclei are present. A considerable difference, moreover, is seen in the greater or less degree of concentration which the glandular masses had undergone. In *Nymphon hamatum* and in *Ascorhynchus glaber* the gland in the fourth joint (which, however, in the latter species is also present in the fifth joint of the leg) forms only a single mass, which runs through the whole joint and opens at the one side in a row of pores, each of which is placed at the tip of a chimney-like process. As seen in fig. 6, the gland in *Nymphon hamatum* almost extends on the one side of the leg, between the intestinal cæcum and the

[1] These glands are also observed by Dohrn; he calls them "ein zweites noch grösseres Excretions- (oder Drüsen?-) Organ (*Ibid.*).

wall, from the one longitudinal ridge to the other; whereas in *Ascorhynchus glaber* (fig. 9) the row of pores is placed almost exactly opposite to the single but very strong ridge.

In *Pallene* and *Phoxichilidium* the structure and the position of the glands show nothing particularly interesting. In *Pallene* there are many pores, in *Phoxichilidium insigne* only a single pore at the end of the joint, placed at the tip of a conical excrescence (Pl. XVI. fig. 18k). The structure of these glands in *Colossendeis* is extremely interesting. In the three species of this genus in which I studied them, *Colossendeis leptorhynchus*, Hoek, *Colossendeis megalonyx*, Hoek, and *Colossendeis proboscidea*, Sab. (sp.), the gland consists of very numerous more or less isolated parts of a rounded or more longitudinal shape, each of them opening separately by a distinct pore, or (*Colossendeis proboscidea*) three or four opening together in a single pore. In fig. 14 on Pl. XVI. I figure a part of the integument of *Colossendeis megalonyx* magnified; in fig. 15 of the same plate a part of the integument of *Colossendeis leptorhynchus* is shown. The glands seem to correspond with a wide vesicle (*o*), in the interior of which a narrow canal (*p*) lies wound spirally; this canal is easily traced till it opens at the pore. Those conical cavities into which the glands are seen penetrating are much wider than the others. For the structure of the gland itself the specimen of *Colossendeis leptorhynchus* which I investigated was not all I could have wished. On a transverse section it looks quite as if all the original glandular cells had dropped from the connective tissue, this tissue itself being the only part that remained as an empty skeleton. Perhaps the state of preservation is to a certain extent the cause of this. The male specimen of *Colossendeis proboscidea*, in which also I studied these glands was in a much better condition. Plate XVIII. fig. 2 shows the distribution of the glands (*g*) over nearly half the inner circumference of the skin of the fourth joint. Moreover, in fig. 3 a small part of this skin is figured more strongly magnified with the glands opening into one of the pores. The gland itself (*g*) shows a dense ball of round and nucleated glandular cells. A comparatively wide and very transparent canal extends from the gland to the interior of one of the integumentary cavities (*c*), and a very narrow duct (*d*), which is irregularly rolled up, runs through this wide canal till it reaches the pore at the end of the cavity. When studying a part of the skin of the leg from the interior it is easily seen that three or four of these glandular bodies send their ducts into the same integumentary cavity. From the beautifully developed net-work of nerves and ganglionic plexuses, which extends over the whole inner surface of the integument, distinct nerves are seen arising and penetrating the cavities or innervating the glands.

Finally, I wish still to mention the curious manner in which these glands of the fourth joint of the male open in *Oorhynchus aucklandiæ*, Hoek. A very long cylindrical appendage is inserted on the fourth joint a little behind the middle. The gland opens at the tip of this appendage by means of a very long duct, which shows a distinct swelling (a kind of receptaculum) near the beginning.[1] About the structure of the gland itself in this species I have no observations to communicate.

2. *Nervous System.*—Of the different systems of the Pycnogonida the one most eagerly studied is, without doubt, the nervous system, and this is quite natural, because it has been rightly considered, that if any system could be expected to shed light on the affinities of the Pycnogonids with the other Arthropoda, it would be the nervous system.

Among the more important papers on the subject, those of Zenker, Semper, Dohrn, and myself may be mentioned. The way in which Zenker (*loc. cit.*) treats of the nervous system of *Nymphon* is not a very happy one, as he describes and figures it as consisting of a supra-œsophageal ganglion and four thoracic ganglia. The account given by Semper[2] of the nervous system of this genus is much more accurate. He tells us that in *Nymphon* the supra-œsophageal ganglion innervates the mandibles and the eyes, and that the first of the five thoracic ganglia furnishes nerves to the proboscis, to the palpi, and to the ovigerous legs, while the four following ganglia give off nerves to the four legs. The number of thoracic ganglia is, according to Semper, also five in *Pallene* and in *Achelia*, on the contrary only four were observed by him in three species of *Phoxichilidium*. In my paper the optic nerves of *Pycnogonum* are described, and the number of ganglia in *Nymphon* is given as five, in *Pycnogonum* as four.[3] We find in Dohrn's latest paper (*loc. cit.*, p. 37) a much more detailed description of the structure of this system. The supra-œsophageal ganglion innervates the mandibles, and, moreover, gives off an azygous nerve, which dorsally innervates the proboscis, and forms a ganglion at about one-third from the extremity of the proboscis. The first thoracic ganglion gives off three pairs of nerves; the first pair arising from the ganglion a little outside and below the insertion of the circum-œsophageal commissures, innervates the lateral parts of the proboscis. Like the azygous proboscideal nerve, they form ganglia at about one-third from the extremity of the proboscis, and these three ganglia are connected by commissures, which thus form a secondary œsophageal ring. The second pair innervates the so-called palpi; the third arises from the ganglion laterally towards the posterior part, it innervates the ovigerous legs. Moreover, Dohrn observed that this first thoracic ganglion not only in the genera furnished with palpi and ovigerous legs, but also in those forms which have lost their palpi and even in the females, which have lost also their ovigerous legs, consists of three nuclei of "fibrillären Punktmasse," each of which gives off the fibres for the nerves respectively of the proboscis, palpi and ovigerous legs. In a young stage of the embryological development, Dohrn made the observation that the first ganglion really consisted of two

[1] Such a long appendage, at the tip of which the gland opens, occurs also in *Ammothea* (Dohrn, *loc. cit.*, p. 36).
[2] Semper (C.), Über Pycnogoniden und ihre in Hydroiden schmarotzenden Larvenformen (Arbeiten a. d. Zool.-Zoot. Institut in Würzburg, i., 1874, p. 278).
[3] *Loc. cit.*, p. 249.

not quite separated pairs of ganglia. Finally, Dohrn states in the same paper that besides the five (six) double ventral ganglia there are two others, which, however, in some genera, totally disappear, and in other genera are preserved only in a rudimentary condition. Accordingly, Dohrn observed immature stages of *Phoxichilus*, in which, behind the sixth ventral ganglion, were present two distinctly separated, although much smaller, pairs of long ovate ganglia. Of these the first pair gives off no nerve, and the second pair the two nerves for the abdomen.

For my observations with regard to the nervous system of the Pycnogonids, I made use of the following specimens from the material of the Challenger Expedition :—one of *Nymphon hamatum*, one of *Nymphon brachyrhynchus*, two of *Nymphon brevicaudatum*, Miers ; numerous specimens of *Nymphon robustum*, Bell ; one of *Colossendeis leptorhynchus*, one of *Colossendeis megalonyx*, and one of *Colossendeis proboscidea*, Sab. ; finally, one of *Phoxichilidium pilosum*. What I tried to ascertain in my investigations was, in the first place, the innervation of the cephalic appendages and of the proboscis, in the second place the structure of the first thoracic ganglion, in the third place that of the last thoracic ganglion, and the presence or absence of the two rudimentary abdominal ganglia. In how far I have been successful in this may be judged from the following :—

The nervous system consists in all species of Pycnogonids of a supra-œsophageal ganglion, an œsophageal commissure, and five (seldom four) thoracic ganglia. The supra-œsophageal ganglion is situated in the cephalothoracic segment; however, its place varies greatly with the form of this segment, and therefore it even shows small differences in the different species of one genus. The differences in the different genera are more considerable. In the genus *Nymphon* it is, as a rule, placed towards the hinder extremity of the cephalic part of the cephalothorax, below the oculiferous tubercle, and above the insertion of the two ovigerous legs. In the genus *Colossendeis* this ganglion is found nearly in the middle of the cephalic part of the cephalothoracic segment, which part is usually separated (distinctly in *Colossendeis leptorhynchus*, e.g.) from the remaining part of the segment by means of a constriction; the oculiferous tubercle is here exactly above the ganglion. In *Phoxichilidium* the ganglion is found also above the insertion of the ovigerous legs; but the oculiferous tubercle is in most species of this genus situated quite anteriorly, on that part of the cephalothoracic segment which overhangs the proboscis. As a rule the shape of this ganglion is round (Pl. XVIII. fig. 6), but in *Colossendeis* the comparatively small ganglion is much broader than long (Pl. XVIII. fig. 4). Whereas, in the other genera the two œsophageal commissures are very short, run parallel to each other, and enclose a narrow canal through which only the œsophagus passes (Pl. XVIII. fig. 11 C.), in *Colossendeis* (I observed it in *Colossendeis leptorhynchus*, Hoek, and in *Colossendeis proboscidea*, Sab., sp.) this canal is very wide (Pl. XVIII. fig. 4), the commissures which connect the supra-œsophageal ganglion with the first thoracic ganglion are very long, and through the wide canal formed by these not only the intestine passes but also two large muscles (Pl. XVIII. fig. 7), which serve for the movement of the proboscis, and run from the posterior margin of the proboscis to the hinder part of the dorsal surface of the cephalic part of the cephalothorax. These muscles are anteriorly narrow, and grow considerably broader towards their extremity, so that their shape is rather triangular. The nerves which arise from the supra-œsophageal ganglion are the same in *Nymphon, Phoxichilidium*, and *Colossendeis*. To the front three comparatively strong nerves are always present.[1] Of these, that nerve which is placed in the middle takes its origin in the ventral surface of the ganglion, a little behind the front margin. This middle nerve is the azygous proboscideal nerve. It runs close to the dorsal surface of the proboscis, and gives off at rather irregular distances extremely small nerves, which innervate the integument. At a distance of about one-fourth of the total length of the proboscis in *Nymphon*, and of about one-eighth of the same length in *Colossendeis*, it enters the ganglion (Pl. XVIII. fig. 8), which, no doubt, has already been observed by Dohrn. The two mandibular (antennary) nerves arise from the front margin of the supra-œsophageal ganglion ; they are nearly (in *Nymphon* and in *Phoxichilidium*) of the same strength as the azygous proboscideal nerve. First they are slightly divergent, then they approach each other again so as almost to touch the azygous proboscideal nerve, then they separate again and direct themselves forwards and a little upwards, running in the mandibles very close to that part of the dorsal surface which is nearest to the mandible of the other side (Pl. XVII. fig. 4, *n.m.*). In regard to the farther course of the mandibular nerve I could only observe that it divides into two branches. These nerves are not totally wanting in the genus *Colossendeis*, and considering the case of *Colossendeis gracilis* with its distinct mandibles, we need not wonder at this. Yet the nerves are very short and represented only by rudimentary branches (Pl. XVIII. fig. 4, *m*). (Those of the interesting specimen of *Colossendeis gracilis* I was unable to observe without dissecting the specimen.)

With regard to the innervation of thee yes, I observed distinct nerves arising from the supra-œsophageal ganglion. These I have described already in my paper on Pycnogonids, published in 1877, and their presence is confirmed by Dohrn's observations. Dohrn (*loc. cit.*, p. 37) says that they arise from the sides of the ganglion, and that they are widely separated from one another. Moreover, Dohrn tells us that there are two of them, and that they divide and innervate the eye in a peculiar way. I studied the innervation of the eyes in *Nymphon brachyrhynchus, N. strömii, N. robustum*, and in *Colossendeis proboscidea*. The latter species is a blind one ; its eyes are represented only by a small

---

[1] In the figure I give of the nerves arising from the supra-œsophageal ganglion of *N. robustum* (Pl. XVII. fig. 4), besides these three nerves two thinner ones are figured. These, however, are not present, as I ascertained after the plate was printed off. Although I have given myself much trouble in trying to determine what it was I had mistaken for nerves, I have not succeeded. The thin threads have the appearance of narrow ducts ; they extend backwards to behind the supraœsophageal ganglion and pass between this ganglion and the upper surface of the œsophagus.

rounded spot distinguishable by its transparency. The optic nerves are represented by two strong bundles of nerves, arising as in *Colossendeis megalonyx* (Pl. XVIII. fig. 4, *o*) laterally from the dorsal anterior surface of the ganglion. These nerves divide into stronger and feebler branches, and the latter have small ganglia between them. Such small ganglia are present also on the interior surface of the small transparent spot representing the rudimentary eye. They are in relation partly with the nerve fibres of the so-called optic nerve, partly with the complicated system of nerves and ganglia, which covers in this species as in the other species of the same genus, the internal surface of the integument. In this species of the genus *Colossendeis*, and from what I have seen of the matter, the same is the case in *Colossendeis leptorhynchus* and *C. megalonyx*, the optic nerve-bundles are really integumentary nerves, giving off the nerve-branches, which, as will be shown hereafter, extend all over the inner surface of the integument, having everywhere ganglia between them, and, as a rule, at small distances from one another.

I feel inclined to consider the condition of the eye, and of its innervation as described for *Colossendeis proboscidea*, as the original condition in the Pycnogonida. As an example of the most highly developed condition, I wish to describe that of *Nymphon brachyrhynchus*. Dorsally the supra-œsophageal ganglion gives off two nerves, which are at their origin quite covered over by ganglion cells (Pl. XVIII. fig. 11 C. *o'*). Where they arise from the ganglion, the distance between the two nerve-bundles is not very considerable; they diverge slightly till they reach the base of the oculiferous tubercle. Here each of these nerve-bundles divides into two branches which run horizontally, the one towards that part of the integument which is before, the other towards that behind the oculiferous tubercle. On reaching the integument each of these two branches divides into numerous smaller nerves and nerve-fibres; moreover, they send upwards numerous nerve-fibres, which penetrate the oculiferous tubercle and extend till they reach the outer wall of the eye. There can be no doubt, therefore, that the eye is innervated by nerve-fibres not united together so as to form a distinct optic nerve.

The same mode of innervating the eye I observed in *Nymphon strömii*. For the study of the innervation of the rudimentary eye of *Nymphon robustum* I got good preparations also. A part of one of these is figured in Plate XVIII. fig. 5. We have the same nerve-bundles covered by ganglionic cells; they do not, however, divide immediately into the stronger integumentary and smaller optic nerves, but before doing so they enter a comparatively large ganglion (fig. 5, *g*) from which these nerves are seen arising. This same ganglion gives off also the nerves for the setæ, which in this species are present in considerable number at the tip of the oculiferous tubercle.[1]

---

[1] This quite corresponds with Dohrn's observations, that the optic nerves give off also branches to the curious organs observed by him between the eyes. I believe these organs are rudimentary in *Nymphon*, in *Colossendeis proboscidea* they are totally wanting.

While the eye of *Colossendeis proboscidea*, Sab., must probably be considered as the most primitive condition, that of *Nymphon robustum* is undoubtedly a degenerated form. In the innervation of the different stages of development of the eye of the Pycnogonids I believe I have pointed out one common feature, viz., that the two nerves arising from the supra-œsophageal ganglion may be only partly considered as optic nerves, and that it is much more in accordance with the facts to call them integumentary nerves, branches of which have assumed the function of optic nerves. For the investigation of the development of the organs of sense and especially of the eyes, I believe the study of the Pycnogonida will yield in future very interesting results.

The number of thoracic ganglia in *Nymphon, Colossendeis,* and *Phoxichilidium pilosum* is five. Those of *Colossendeis* (*C. proboscidea*, Sab., sp.) are figured in Plate XVII. fig. 2; those of *Nymphon robustum*, Bell, in fig. 3 on the same plate. In these three species the first ganglion (better called the first ganglionic mass) is separated from the second by two distinct commissures, and in *Nymphon robustum* and *Phoxichilidium pilosum* even the outward form of the ganglion shows its complex nature; on a longitudinal section it is seen to be composed of two distinct ganglia. This anterior ganglionic mass supplies the two ventral parts of the proboscis, the palpi, and the ovigerous legs; three pairs of strong nerves are given off by the ganglion, which, judging from its anatomical structure and from its development (*vide* Dohrn, *loc. cit.*, p. 34, and also in this report *sub* embryology, with Pl. XIX. figs. 11 and 13) is composed of only two pairs of original ganglia.[1] For a long time I was greatly puzzled with this fact, until the study of the nervous system of *Colossendeis* dispelled my doubts.

In fig. 4 of Plate XVIII. I figure the supra-œsophageal and first thoracic ganglia of *Colossendeis megalonyx*. The latter gives off the two nerves for the ovigerous legs (*o*), the nerves for the palpi (*pa*), which in the same way as the nerves for the legs immediately divide into two branches, and in the third place the two nerves for the proboscis (*t″*). Moreover, a fourth nerve (*t′*) is observed, which serves also for the innervation of the proboscis, and the fibres of which arise from the same part of the ganglion as those of the main proboscideal nerve. For a short way these fibres run parallel with the fibres of the commissures, so that this first pair of proboscideal nerves seems to arise from these commissures.[2] These same nerves are also present in

---

[1] On page 32 of the same paper, Dohrn asserts that in the first thoracic ganglion three nuclei are present of the well-known "fibrillären Punktmasse." This does not agree with what I have seen in the three genera I studied, nor does it agree, I believe, with what Dohrn himself says on page 34.

[2] In fig. 2 on Plate XVII. the ventral part of the nervous system of *Colossendeis proboscidea*, Sab., sp., is figured. From the first thoracic ganglion arise the nerves for the ovigerous legs (*n. o. l.*), and a strong nerve (the palpus nerve) dividing into two branches (*n. pr.* and *n. p.*) The most anterior, and at the same time most dorsal, part of this ganglion, from which arise the two pairs of proboscideal nerves, and the commissures, has not been figured, having been removed during the preparation.

Nymphon in a much more rudimentary state. Nobody would conclude, however, from their presence that the first ganglion was originally composed of four ganglia; but the supposition of its being formed of three nuclei loses at the same time much of its value, and the ventral part of the proboscis must be considered as being innervated by the same original ganglion as that which gives off the nerves for the palpi.

I believe there can be no doubt that we have here the original condition of the nervous system; at the same time this fact may be considered as suggesting the opinion that the palpi originally belonged to a pair of appendages which coalesced to form the two undermost of the three parts of which the proboscis is composed.

The two front nerves of the first thoracic ganglion in *Nymphon* and *Phoxichilidium*, and the strongest of the two front nerves of the same ganglion in *Colossendeis*, enter the proboscis and run forwards exactly in the middle of the two ventral parts of the proboscis, which I compared (note on p. 14) with the carpels of a monocotyledonous fruit. These nerves I call the paired proboscideal nerves. They end, like the azygous proboscideal nerve, by entering a ganglion, placed at about the same distance from the end of the proboscis as the ganglion of the azygous proboscideal nerve. These three ganglia are united by a ring, which runs between the outer wall of the proboscis and the chitinous wall of the œsophagus, among the numerous muscles which run from the one wall to the other.

So far my description quite agrees with that of Dohrn, as given above. However, a considerable difference arises from the fact that the ganglia which were seen by Dohrn are not to be considered as ganglia of the azygous or paired proboscideal nerves, but as being really the terminal ganglia of three strong nerve bundles, composed of nerve fibres and ganglia, which run longitudinally below or above the three stout proboscideal nerves, so that they lie between these nerves and the wall of the œsophagus. The discovery of these three ganglionic nerve bundles has been very fortunate. It is curious that they have hitherto been always overlooked, and especially that Dohrn did not observe them. But then it must be considered that these nerve bundles are placed among numerous muscles running over and beneath them, and making a preparation totally impossible. A successful longitudinal section, made exactly above or below a bundle, is the only way to detect it. I call these bundles ganglionic, for although I do not believe that their function is analogous with that of the sympathetic system of higher animals, yet their structure shows in general the same relative distribution of ganglion cells and nerve fibres as in the case of the ganglionic system of higher animals.

Fig. 6 on Plate XVIII. shows the position of these nerves in the proboscis; while fig. 8 shows a part of one of them more strongly magnified. Each of them (*g*) consists of a strong bundle of nerve fibres, which, posteriorly at irregular, anteriorly at more regular distances, are surrounded by groups of ganglion cells. Thus each of the cords has the appearance of a row of ganglia connected by bundles of nerve fibres. The size of these ganglia is not quite the same over the whole cord, the foremost being slightly larger than those placed more posteriorly. As to their shape, I observed the following two different types. In some of the ganglia the cells are placed on both sides of the bundle, which passes through it, and these ganglia have a very regular rhomboidal form. The other type is represented by those ganglia in which ganglion cells are to be observed only on one side of the nerve bundle, and which accordingly show a triangular form. The triangular ganglia seem to be more numerous in *Colossendeis*, the rhomboidal form in *Nymphon*; in both genera, however, the stout ganglia, which are placed in the front part of the cord, and in the first place the comparatively large ganglion (figs. 6 and 8 *y*) observed by Dohrn are of a distinct rhomboidal form.

The form of the ganglion is, of course, determined by the number of nerves which branch off from it. The different ganglia give off besides numerous smaller nerves, one (in the triangular form) or two (in the rhomboidal) stouter nerves. These run in the foremost part from the one ganglionic bundle to the two others, and form nerve-rings (fig. 6, a^{I}, a^{II}, a^{III}, &c.), of which the secondary œsophageal ring (figs. 6 and 8 *a*) observed by Dohrn is the first and the stoutest. In *Nymphon robustum* I observed five or six of these nerve rings, but in *Colossendeis* they are still more numerous.

With regard to the three stout proboscideal nerves, which have been observed already by Semper and Dohrn, and which, according to the latter author, terminate in the three ganglia (the front ganglia of my ganglionic bundles), I have ascertained that they are connected with the ganglionic bundles in the following way :—They run superiorly to and quite independently of the ganglionic bundle, till they reach the last but one ganglion of that bundle (fig. 8 *u*). This they enter, their fibres passing through it and contributing to the comparative thickness of that part of the bundle which unites the last but one and the last of the ganglia (fig. 8 *y*). However, it is very probable that at least some of these fibres extend beyond the last of the ganglia. I am not quite certain whether perhaps, a union of the ganglionic cord with the proboscideal nerves does not also take place posteriorly. As I have stated already above, the proboscideal nerve gives off branches ; and about the middle of the proboscis of *Colossendeis proboscidea*, Sab. (sp.), on both sides of the stouter middle nerve two thinner cords run parallel with it in its immediate neighbourhood ; these are branches of the middle nerve. Investigating the first part of these lateral branches, close to their origin from the main proboscideal nerve, I once observed (in *Colossendeis megalonyx*, Hoek) very small ganglia with thin nerve threads running along this nerve without, however, exchanging fibres ; these are, possibly, the end branches of the ganglionic bundle. While the origin of these branches and their connection with the proboscideal nerves is so easily noticed, with regard to their termination I only observed that the bundles, when approaching the end of the proboscis, become thinner and thinner, the ganglia smaller, and the lateral branches arising from the ganglia more numerous and much thinner and shorter. This is quite in correspondence with the number of muscles which these ganglia innervate; whereas these in the front part of the proboscis are stouter and separated by distinct longitudinal intervals, in the posterior part they are thinner and placed almost in an uninterrupted row.

This brings us quite methodically to the function of these ganglionic nerve-bundles. In the vertebrates we can distinguish by the microscopical structure sympathetic ganglia and nerves from those of the cerebro-spinal system, but in the invertebrata this is by no means so easy. In the first place, we must consider the function of the part of the nervous system in question. The ganglia and the nerves of my three ganglionic bundles innervate the striped muscles of the proboscis. Unstriped muscular fibres are by no means rare in the muscular tissue of the Pycnogonids,[1] but even if they were quite wanting, as they seem to be in the muscular tissue of the Crayfish,[2] those of the proboscis ought to be considered as voluntary fibres. Moreover, the action which the food undergoes in the proboscis by means of these fibres is of a purely mechanical nature. Chemical action does not take place in it, therefore comparison of these ganglionic bundles with the sympathetic system of higher animals is impossible.

The morphological explanation of their presence is by no means so easily given. The following reasoning must be considered as an attempt only. The proboscis of the Pycnogonida in the form in which it presents itself should not, of course, be considered as a new organ, only present in this class of Arthropoda. It is only an organ or a combination of organs under a new form, modified under the influence of surrounding conditions. Considering for a moment the supposition right, that it results from the union of three parts, an azygous one placed partly above and partly before the mouth (the upper lip), and two others placed below and behind the mouth, the manducating parts of the mandibles, of which the palpi in that case may be considered as the feelers; then we have in the three nerves, the first of which is given off by the supra-œsophageal ganglion, and the two others arising from the first thoracic ganglion, the normal nerves for the innervation of these parts. With the union of these parts to form a proboscis (and I believe this argument will hold good also if we prefer another homology for these parts), and the predominance of the manducating function of this proboscis, evidently quite a new part of the nervous system, will make its appearance; and it is not difficult to imagine its probable origin.

In the chitinous wall which lines the canal of the proboscis, and which is furnished with rows of very numerous teeth and spines, we have, no doubt (morphologically), a continuation of the integument, so that its inner surface corresponds with the outer surface of the body, while its outer surface, to which the muscles are attached, is the homologue

[1] E.g., in the wall of the vasa efferentia of the males, &c.   [2] Huxley, The Crayfish, p. 181.

of the inner surface of the skin.   Now, this inner surface of the integument both of the body and the legs, especially in species of the genus *Colossendeis*, is richly furnished with ganglia, which spread all over it and are connected with nerves.   They are so very numerous as to form a continuous network of ganglia and nerves, their function being, I believe, to innervate the cavities of the integument, for which I have suggested a respiratory function.   There can therefore be no difficulty in supposing that the ganglia of my ganglionic bundles are derived from originally integumentary ganglia, and that their high development is to be attributed to the changed functions of the parts which surround the mouth.

While these same ganglionic bundles, in a more or less developed state, are found in all species and genera of Pycnogonida, it is very probable, I believe, that in the other classes of the Arthropoda their homologues will be sought for in vain.   The shape of the terminal ganglia, of which the dorsal one is the largest, is best seen from the drawing (Pl. XVIII. fig. 8).   Of the nerves which arise from it, two run in an oblique direction (one to each side), these enter again (at least in *Nymphon*) a small ganglion, from which nerves are given off to the tactile organs placed in the so-called lips of the proboscis. Of these small ganglia, those two, which are placed on both sides of one of the lines of union of the three proboscideal parts, are again connected by means of a nerve string. The tactile organs consist of a small tuft of hairs placed just at the end of the chitinous list which marks the place of union of two of the proboscideal parts meeting there laterally.   Perhaps the nerve fibres of the small nerve bundles, which enter the secondary ganglia and innervate these tactile hair-tufts, take their origin in the three original proboscideal nerves.

Besides the three original nerves and the three ganglionic bundles, two thinner nerves enter the proboscis dorsally.   These I observed only in *Nymphon* arising from the supra-œsophageal ganglion.   The two thin nerves which in *Colossendeis* run alongside and quite near to the main proboscideal nerve must be considered as branches of this main nerve, and no doubt there are still other longitudinal nerves, which run through the proboscis, and which must also be considered as branches of one of the three main nerves.

What I observed in regard to the remaining part of the nervous system is the following :—The shape of the four thoracic ganglia may be seen from the figures on Plate XVII.   The length of the commissures uniting these ganglia is different in different genera, and even in the different species of one genus.   In *Nymphon robustum*,

---

[1] It seems to me that an analogous case is that of the visceral or stomatogastric nerves of the Crayfish, studied by different authors, and investigated recently more accurately by Prof. Huxley (Anatomy of Invertebrated Animals, London, 1877, p. 330), a complex nervous apparatus, serving chiefly for the innervation of the muscles of the mandibles, and for that part of the intestine which has been called by Huxley the gastric mill.   This gastric mill of the Decapod Crustacea is placed behind the œsophageal commissures ; the analogous apparatus of the Pycnogonids is found in front of the same commissures.

Bell, *e.g.*, the commissures between the third and fourth ganglia are the longest; in *Nymphon brachyrhynchus*, however, those between the second and third are the longest. In *Colossendeis* and in *Phoxichilidium pilosum* the relative length is the same as in *Nymphon robustum*. In all the species and genera I observed the commissures between the last two thoracic ganglia to be shorter, often much shorter, than those between the preceding ganglia. The nerves which arise from these ganglia in *Colossendeis* immediately divide into two branches, whereas in *Nymphon* they only separate after having reached the middle of the lateral process at the end of which the leg is inserted. In *Phoxichilidium pilosum* they again divide immediately after leaving the ganglion. The last ganglion has a truncated form; in most of the genera I studied I observed at least a trace of the presence of the small accessory ganglia (abdominal ganglia) which Dohrn observed in *Phoxichilus*, and which I found beautifully developed in a young specimen of *Colossendeis proboscidea*.

Close to the hinder margin of the ganglion two small excrescences arise from its dorsal surface. They are separated from one another by a small space, and as they unite again at a short distance from the ganglion a round area is left between them. In a full-grown specimen of the same species only a single excrescence was observed, arising from the hinder margin, and I observed it of the same form in *Colossendeis megalonyx* and also in *C. leptorhynchus*. I did not observe it in *Nymphon*; but in the species of that genus which I studied (*N. strömii* and *N. robustum*) I always saw two distinct medullary nuclei in the interior of the ganglion placed close to its hinder margin, behind the large medullary masses from which the nerves for the hindmost pair of legs arise.

According to Dohrn, the nerves which arise from this ganglion are two in number, besides the two stout nerves for the hindmost pair of legs. It is extremely difficult to avoid confounding threads of connective tissue, arising from the neurilemma that surrounds this ganglion with nerves. I have, however, distinctly observed that often four different nerves branch off from the ganglion; in *Colossendeis proboscidea*, *e.g.*, two smaller ones are given off more in the middle from the dorsal excrescence, and two stouter ones leave the ganglion nearer to the nerves which innervate the fourth pair of legs. Two pairs of nerves arising from the hinder surface are also present in *C. leptorhynchus*; but, strange enough, only one pair (that arising from the excrescence in the middle of the hinder margin) in *Colossendeis megalonyx*. In the genus *Nymphon* the two nerves which arise from the ganglion leave it as one single nerve, and only afterwards divide into two branches. The same difficulty is experienced in ascertaining the presence of nerves branching from the commissures which unite the different thoracic ganglia. These are present, however, in all the genera, and even in all the species I studied. As a rule, they leave the commissures much nearer to the preceding than to the following ganglion, but often also nearly in the middle of the commissures; between the second and third, between the third and fourth, and between the fourth and fifth thoracic ganglion, distinct nerves are observed arising at an acute angle from the commissures. These innervate the muscles which run from the one segment to the following one.

I give a figure (Plate XVIII. fig. 10) of one of the small ganglia and the nerve fibres arising from it on the inner surface of the integument. I observed these ganglia in *Nymphon* and in *Colossendeis;* they are more numerous in the latter genus, and especially numerous and in a better state of preservation in one of the specimens of *Colossendeis proboscidea,* Sab., dredged last summer north of Scotland. The figure is taken from a part of the integument of a specimen of this species.

To show the minute structure of the ganglia I give in fig. 11 A–F a series of six figures illustrating vertical (frontal) sections through the supra-œsophageal and first thoracic ganglia. As will be easily seen, the sections do not form a complete series, they are only the most interesting out of a series of about twenty. Fig. A represents a section quite at the front of the supra-œsophageal ganglion; a distinct and comparatively thick neurilemma sheath surrounds the whole ganglion, and three distinct medullary nuclei (pointed substance according to Leydig) show the place of origin of the three main nerves. The rest of the section is composed of ganglion cells, with the exception of a small medullary spot at the left side, which does not occur at the other side, and proves that the section is somewhat oblique. Neither the œsophageal commissures nor the first thoracic ganglion are yet to be seen in this section. Fig. B represents a section which passes through the supra-œsophageal ganglion and through the foremost part of the first thoracic ganglion, but not yet through the commissures. Almost the whole section is occupied by the medulla, which forms regular prominences towards the periphery, and only a very small space is occupied by the ganglion cells; *m* is the lobe for the mandibular (antennary) nerve, *o* represents the lobe for the optic nerve, and *x* is a median lobe about the function of which I do not feel quite sure. In section C this median lobe is no longer to be observed, and the optic lobes have here assumed a much more elongated form. In section B the commissures are still totally wanting, but the front part of the first thoracic ganglion, with the medullary nuclei (*p*) for the two infra-proboscideal nerves, are distinct. These in section C are reduced to small lobes, while those for the nerves of the ovigerous legs (*t*) are distinct. In this section the œsophageal commissures are seen, and also the undermost parts of the two optic nerves, which arise from the supra-œsophageal ganglion (*o'*). Their connection with the optic lobes, however, does not occur in this section. Section C passes almost through the middle of the oculiferous tubercle, so that two eyes are placed in front of this section and the two others behind it. Section D represents the last part of the supra-œsophageal ganglion; the hindmost part of the medulla is seen, and the ganglion cells begin again to increase in number. Of the first thoracic ganglion, the section goes through that part of the medullary centrum which gives off the nerves for the ovigerous legs (*l*). In E this medullary centrum is considerably less voluminous, and, finally, in F only two rounded spots of medulla are left (co'), which give off the commissures between this and the second thoracic ganglion.

About the histological structure of the ganglia I wish to be very short. In fig. 12 on Plate XVIII. a longitudinal section is given of the first thoracic ganglion of *Nymphon brachyrhynchus*. A comparatively thick sheet of connective tissue (neurilemma sheath) surrounds the ganglion, and is in continuation with the sheath of the commissure which unites this ganglion with the second thoracic ganglion. Small ganglion cells, each with a distinct nucleus, fill that part of the ganglion which is not occupied by the nerve fibres. These ganglion cells are situated in meshes of connective tissue, they are small and furnished with but little protoplasm; their nuclei are distinct, and show a small and glittering nucleolus.

In a section through the first thoracic ganglion of *Colossendeis leptorhynchus*, it is easily observed, that the ganglion cells are of two different sizes: there are very large ones rich in protoplasm, and furnished with comparatively large nuclei; there are also very small ones, which show only a small quantity of protoplasm. Fibres and sheets of connective tissue are everywhere observed between the ganglion cells; the neurilemma sheath of the ganglion itself is very thick, whether it is to be considered as really double (an outer and an inner neurilemma sheath, the latter of which should be in connection with the connective tissue meshwork of the interior of the ganglion), I have not been able to ascertain.[1]

3. *Eyes.*—Of the organs of sense I paid special attention to the eyes. Besides the tactile organs spread over the whole surface of the body, the curiously shaped hairs of the palpi of *Ascorhynchus*, and perhaps of other genera, which probably are also organs of a special sense (olfactory?), the eyes are the only certainly known organs of sense in Pycnognonids.[2] In the first place, I tried to ascertain in how far the eyes can really be said to disappear in those animals which inhabit great depths. I therefore drew up the following list, in which the species from the "Challenger" and "Knight Errant" expeditions are placed according to the depths they inhabit.

---

[1] Hitherto I have studied only such sections of the ganglia as are made by cutting the whole body. These are sufficient for researches on the general anatomy, but totally insufficient with regard to histology. The numerous sheets and strips of connective tissue prevent the substance used for enclosing the body (paraffine or Calberla's substance) from penetrating the whole interior of the body. Thus some parts always remain disunited, and the cutting of thin sections is extremely difficult, if not impossible.

[2] The curious organ which Dohrn observed between the two eyes on each side of the oculiferous tubercle I have not observed. Perhaps it does not occur in *Nymphon*, the only genus in which I investigated the eyes.

Name of Species.	Depth in Fathoms.	Number of Eyes.	Condition of the Eyes.	Observations.
Discoarachne brevipes, Hoek, .	...	Four.	Small; placed on a blunt tubercle.	
Hannonia typica, Hoek, .	...	"	Small; placed on a rounded tubercle.	
Phoxichilidium fluminense, Kröyer,	7–20	"	Two anterior a little larger than the two posterior.	
Phoxichilidium insigne, Hoek, .	7–20	"	Size of the eyes different, as in the foregoing.	
Nymphon brachyrhynchus, Hoek, .	10–120	"	Large; oculiferous tubercle pointed.	
Nymphon brevicaudatum, Miers, .	25–125	"	Small; oculiferous tubercle extremely elevated.	
Nymphon fuscum, Hoek,	25	"	Small; touching one another at the tip of a not very prominent tubercle.	
Ascorhynchus minutus, Hoek, .	38	...	...	A blunt oculiferous tubercle, without eyes or pigment.
Pallene languida, Hoek, .	38	Four.	Small; oculiferous tubercle not very prominent.	
Pallene lævis, Hoek, . .	38–40	"	Middle-sized; tubercle blunt and low.	
Pallene australiensis, Hoek, .	38–120	"	Middle-sized; tubercle acute.	
Phoxichilidium patagonicum, Hoek,	45–120	"	Two anterior large, the two posterior much smaller. Tubercle high, but blunt.	Form of the tubercle differs with age.
Pycnogonum litorale, Ström, .	53	"	Extremely small; tubercle blunt.	
Colossendeis megalonyx, Hoek, .	55–120	"	Two anterior a little larger than the two posterior. Tubercle elevated.	Form of the tubercle differs with age.
Nymphon grossipes, Oth. Fabr., sp.,	83–540	"	Large; placed at the base of a high and acute oculiferous tubercle.	
Nymphon brevicollum, Hoek, .	83	"	Large; oculiferous tubercle low.	Oculiferous tubercle obliquely truncated.
Colossendeis robusta, Hoek,	120	"	Two anterior large, two posterior small; oculiferous tubercle elevated and acute.	
Ascorhynchus orthorhynchus, Hoek,	150	"	Middle-sized, with pigment; oculiferous tubercle acutely pointed.	
Nymphon robustum, Bell, .	375–540	...	No lens, and no pigment.	
Colossendeis leptorhynchus, Hoek, .	400–1600	...	Two small spots (rudimentary lenses?) at both sides of the blunt tubercle.	Younger specimens with a more elevated tubercle.
Nymphon strömii, Kröyer, .	515–540	Four.	Two larger ones directed forward, two smaller ones backward. Tubercle blunt.	
Nymphon macronyx, G. O. Sars,	540	"	Not very large; tubercle bifid at the extremity.	
Colossendeis proboscidea, Sab. (sp.), .	540	...	Two small spots (rudimentary lenses?) on both sides of the oculiferous tubercle.	Tubercle of the younger specimens elevated.
Phoxichilidium patagonicum, var. *elegans*,	600	Four.	Two large eyes directed forward, two smaller ones directed backward. Tubercle high, acute.	
Oorhynchus aucklandiæ, Hoek, .	700	"	Small, with pigment; tubercle directed forwards.	
Nymphon perlucidum, Hoek, .	825	...	...	Oculiferous tubercle low, without pigment.
Nymphon longicoxa, Hoek, .	1100	Four.	Small; the pigment light-brownish; tubercle low and blunt.	
Nymphon compactum, Hoek, .	1100	...	...	Oculiferous tubercle represented by a round spot.
Colossendeis minuta, Hoek, .	1250	...	...	Two small rudimentary spots, without pigment; tubercle blunt.
Ascorhynchus glaber, Hoek, .	1375	...	...	Oculiferous tubercle elevated.

Name of Species.	Depth in Fathoms.	Number of Eyes.	Condition of the Eyes.	Observations.
Nymphon hamatum, Hoek,	1375–1600	...	...	Oculiferous tubercle also rudimentary.
Colossendeis gigas, Hoek,	1375–1600	...	...	Oculiferous tubercle of the younger specimens higher, and furnished with two rudimentary spots.
Colossendeis gracilis, Hoek,	1375–1600	...	...	Oculiferous tubercle much elevated, with four lenses.
Phoxichilidium pilosum, Hoek,	1600–1950	Four.	Two large ones directed forward, two very small ones backward; tubercle elevated and acute.	
Nymphon meridionale, Hoek,	1675	„	Distinctly pigmented; tubercle not very high.	
Phoxichilidium oscitans, Hoek,	1675	„	Two very large ones and two small ones.	
Phoxichilidium mollissimum, Hoek,	1875	„	Two extremely large ones, kidney-shaped, directed forward; two very small ones directed backward.	
Nymphon procerum, Hoek,	2160	...	...	Two small spots are present on the tip of a low tubercle.
Nymphon longicollum, Hoek,	2225	...	...	Oculiferous tubercle high, with two rudimentary spots without pigment.
Colossendeis media, Hoek,	2225	...	...	Oculiferous tubercle very high and acute.
Colossendeis brevipes, Hoek,	2650	...	...	Oculiferous tubercle high and acutely pointed.

What we learn from this list is that it is a common feature for the shallow-water species to have four distinct eyes; and for those inhabiting a depth exceeding 400 fathoms to have no eyes, or to have rudimentary eyes without pigment. While exceptions to this rule are rare in the shallow-water species, they are by no means unfrequent in the deep-sea species. As to the structure of those eyes which may be called rudimentary, they often have a distinct lens—a rounded spot marked by its brightness; they are quite destitute of pigment, and, as I learned from a section of the oculiferous tubercle of *Nymphon robustum*, Bell, the small eye has no retina, and is filled up with a mass of detached connective tissue.

In those species which are furnished with distinct eyes, the size of those on the same tubercle is not always the same. In *Nymphon strömii*, Kröyer, the eyes which are directed forwards are slightly larger than those which are directed backwards, but this difference is much more marked in the species of *Phoxichilidium*. This difference in size —as is generally known—is a common occurrence. Grenacher[1] has shown in the simple eyes of the Arachnida and also of the larvæ of many insects, that this difference in size is often accompanied by extremely interesting differences in the minute structure. It

[1] Grenacher, H.—Untersuchungen über das Sehorgan der Arthropoden, 4o, Göttingen, 1879.

would be of interest to study this question in the Pycnogonida, but the limited number of the specimens of the different species of *Phoxichilidium* in the Challenger collection did not allow me to study their eyes.   I investigated those of *Nymphon strömii*, but there the difference in size is slight; I did not ascertain any difference in their structure.

The knowledge of the minute structure of the eyes of the Pycnogonids is of very recent date. They were always considered as simple eyes, and were even mentioned as such by Cavanna. Dohrn, therefore, in the preliminary publication on the results of his studies on Pycnogonids, is the first who gives us some information on these organs.   According to him the eyes, taken in a vertical position, are of a pointed oval shape;[1] they have a retina composed of modified epithelium cells (hypodermic cells), the extremities of which are surrounded by a brown pigment; the cuticula forms a lens.

Among the latest investigations into the structure of the eyes of the Arthropoda, those of Grenacher, published in his splendid memoir,[2] have the merit in the first place not only in proposing a homology between the parts composing a compound and those composing a simple eye, but also in showing the existence of this homology throughout almost the whole type of the Arthropods. Moreover, the value of his monograph with regard to the physiology of the organ of sight, and the immense quantity of special information which it contains, is at present almost inestimable. In investigating the eyes of the Pycnogonids, I was extremely desirous to try whether the scheme for the eye of the Arthropods, as proposed by him, would hold good also in the case of the Pycnogonids. Though my researches did not give me a complete idea of the anatomy of their eyes, I think my results are worth publishing here, because they admit of comparison with the scheme given by Grenacher.[3]

I made numerous preparations of the eyes of different species of the genus *Nymphon*; of *N. brevicaudatum*, Miers, *N. brachyrhynchus*, Hoek, *N. strömii*, Kröyer, and of the rudimentary eyes of *Nymphon robustum*, Bell. I studied more especially those I made of *Nymphon brachyrhynchus* and of *Nymphon strömii*. While the eyes of *Nymphon strömii* (Pl. XVIII. fig. 11, B) are comparatively small, and placed on the sides of a conical tubercle, so as to be separated by a greater distance from one another below than above ; those of *Nymphon brachyrhynchus* are larger, and are placed on the inside of a cylindrical oculiferous tubercle, so close to one another as to meet at their inner surfaces. In fig. 2,

---

[1] Von oben nach unten spitz oval (*loc. cit.*, p. 37).

[2] Grenacher, H.—Untersuchungen über das Sehorgan der Arthropoden, 4°, Göttingen, 1879.

[3] For studying the histology of the Arthropod's eye fresh material, in the first place, is necessary. That I got preparations fit to be used of the eyes of *Nymphon brachyrhynchus*, collected at Kerguelen Island, and put in spirits in January 1874, is almost more than could be expected. In one respect only did the condition of the material hinder me, viz., that I could not make use of some reagents, *e.g.*, of nitric acid. Though used of different strengths, I never obtained the solution of the pigment without the visual rods being destroyed also. Consequently I never got a preparation showing the connection between the fibres of the optic nerve and these rods.

Plate XXI., I figure a longitudinal section of one of the eyes of this species; and in the same figure the place occupied by the outer surface of the eye of the other side is indicated (*a*). This outer surface is invested by a thin chitinous cuticle, which is minutely (longitudinally) striped; round the circumference it is in connection with the chitinous skin of the animal. The inside of this wall is covered with pigment, and it is in this bed of pigment that the union of the fibres of the optic nerve with the elements of the retina takes place. I have already spoken above about the manner in which the optic nerve penetrates the eye. I sometimes got preparations, which made me believe that the optic nerve reached the outer integument of the oculiferous tubercle, at a considerable distance below the eyes, and that it ran along this wall and penetrated the eye where it is in connection with this integument; this, however, is not really the case. Numerous separate nerve fibres reach the under surface of the eye; they then penetrate its cuticle, and in one of my preparations I observed distinctly, that they are in direct connection with the rods of the retina.

In the interior of the eye there is a retina, but there is no distinct vitreous body. I observed that under the thickened cuticula which forms the lens of the eye, just as everywhere else, the epithelium (hypodermis) of the cuticula is formed of rudimentary cells represented by nuclei. Of a distinct row of vitreous body cells, like those figured and described by Grenacher (*loc. cit.*) and Graber,[1] I observed nothing. Nor do I believe that Dohrn's assertion is right, that in the eyes of the Pycnogonida the retina is derived from the epithelium cells (hypodermic cells). This retina consists of rod-forming elements and of ganglion cells. Those parts of the rods which are directed towards the fibres of the optic nerve are thicker and terminate abruptly (Pl. XXI. fig. 4), bearing at the other extremity a long and filamentary appendage; while in other preparations the retina elements which I isolated show the form figured in Plate XXI. fig. 5,—viz., rods which imperceptibly pass into thread-like appendages. In these retinal elements I failed to observe any trace of the presence of nuclei. The thread-like appendages of the rods extend till they reach the cuticular lens. A præretinal lamella, which, according to Graber, is characteristic of the stemma of the tracheate Arthropods is not visible. Between the rod-forming elements numerous ganglion cells are observed in the form of distinct rounded cells. I did not observe whether or not they were really in connection with the filamentary appendages of the rods; but having isolated from one of the preparations the rods, as figured in fig. 5, I found that the rod passes into the filamentary appendage. Probably the ganglion cells have also filamentary appendages, which pass between the filaments of the rods. In this respect therefore my opinion is intermediate to those of Grenacher and Graber. This structure is observed when the section passes vertically through about the middle of the eye

<hr>

[1] Graber, V.—Ueber das unicorneale Tracheaten-und speciell das Arachnoideen-und Myriopoden-Auge. Archiv. f. Mikr. Anatomie. XVII. 1880, p. 58–93, Pl. v.–vii.

(Pl. XXI. fig. 3).   When the same (tangential) section passes through the lateral part
of the eye another structure is observed (Pl. XXI. fig. 2), viz., a mass of reticular tissue,
with numerous ganglion cells in its meshes, not unlike but a little smaller than those
between the rods of the retina.   Empty cavities, from which, perhaps, ganglion cells
have dropped, are observed in it.   A horizontal section of one of the eyes of *Nymphon
strömii* (Pl. XXI. fig. 1) shows the arrangement of these parts in the interior of the
eye.   The middle part is that occupied by the retinal rods, which here have large
ganglion cells at their extremities, and a distinct mass of reticular tissue, separated
from the retinal rods by a thin membrane (?), is observed laterally.   The dimensions of the ganglion cells placed in the meshes of this reticular tissue in *Nymphon
brachyrhynchus* are almost the same as those observed between the retinal rods ; but in
Nymphon strömii the ganglion cells which are found at the extremities of the retinal rods
are much larger.   In regard to the minute structure of these ganglion cells I have hardly
any observations worthy of being mentioned.   The cells are always furnished with a
distinct nucleus, and their contents, as a rule, are granular.   Those I observed between
the rods of *Nymphon brachyrhynchus* are furnished with one filament directed towards
the lens.

The conclusions I have arrived at with regard to the anatomy of the organs of vision
in the Challenger Pycnogonids may be summarised as follows :—

(1.) A vitreous body, developed out of the cuticular epithelium (hypodermis), does
not exist.

(2.) The epithelium cells are present under the lens of the eyes in the same
condition as under the cuticula in general.

(3.) The retina consists of distinct rods and ganglion cells.   Numerous ganglion cells,
placed in the meshes of a reticular tissue, form a mass, which encloses the interior of the
eye occupied by the rods.

(4.) The ends of the retinal rods reach the cuticular lens.   A præretinal interlamella
seems not to exist.

(5.) The retinal rods can not be considered as having resulted from transformed
hypodermic cells.

(6.) The retinal rods have two parts—an innermost thicker part and a filamentary
appendage.

(7.) The eye is surrounded by a chitinous cuticula.

4. *Alimentary Canal and its Appendages.*—I only occasionally got preparations of
the alimentary canal; therefore what I have observed in regard to this organ is far
from exhaustive.

Physiologically, this is perhaps the most interesting organ of all, and, morphologically, its signification is by no means small, as its structure has, along with the number

of the legs, always been considered as a very important support for the belief in the near relationship between the Pycnogonida and the Arachnida, and more especially the Phalangida.

Morphologically, the œsophagus extends from the mouth to behind the œsophageal commissures. Taking the function of the organs into consideration, I believe that only an extremely small part should bear the name of œsophagus. At a very short distance from the mouth the œsophagus widens considerably. This widened part, which shows its greatest dilatation in front of the middle of the proboscis, slopes again backwards, and imperceptibly passes over into a much narrower canal, which extends immediately behind the œsophageal commissures. The widened part of the œsophagus, which almost · reaches to the end of the proboscis, is invested by a chitinous wall. This wall is beautifully beset with thin parallel chitinous bands, which are furnished with numerous thin spines. In the front part these are wanting. They begin on the two ventral parts of the inside a little before they are found on the dorsal part. These foremost spines have the form of short teeth, and only further back do they assume the form of long thin spines or needles. From the wall of this part of the œsophagus numerous bundles of transversely striated muscles extend till they reach the outer wall of the proboscis, their distribution being in *Nymphon*, e.g., such that two longitudinal rows are attached to each of the three parts of which the inner wall of the œsophagus is composed (Pl. XVIII. fig. 9). As to the function of this part of the œsophagus, judging from these muscles and from its internal armature, I think it not very hazardous to compare it with the cardiac portion of the stomach of the Crayfish. It is a masticating apparatus.

Posteriorly it passes over into a very long (slender species of *Nymphon*), or rather short (*Colossendeis*) cylindrical tube, the wall of which is still divided into three longitudinal parts, which on a transverse section are triangular and leave an extremely narrow canal in their middle. I studied the histological structure of this part of the wall, which extends to beyond the œsophageal commissures. Its cells are of a long cylindrical form, longer in the middle and shorter on both sides of the triangular part. They are furnished with distinct nuclei, which sometimes are all placed near the outwardly directed extremity of these cells, but sometimes also are found more in the middle. Between these cylindrical cells there are some of a long conical shape, the base of the cones being, as a rule, directed outwards. Inside, the surface of these cells is invested by a structureless membrana intima, and outside a similar cuticular formation is present (Pl. XXI. fig. 6). This epithelial covering does not end abruptly immediately behind the œsophageal commissures. In the interior of the succeeding part of the intestine it forms three glandular bodies, which hitherto have not been observed, and whose function, judging from their position, must be, I believe, pancreatic. In fig. 7 on Plate XXI. I show the place occupied by these glands, and in fig. 8 of the same plate a transverse section near the extremity of the two undermost glands is figured. These glands project into the interior of the intestine, and are, as far as I know, the only true glandular bodies which stand in connection with the alimentary canal. They are invested by the same membrana intima as the wall of that part of the œsophagus, at the end of which they are found. The form of the cells which compose them is nearly the same. The whole of the gland must be considered as having taken its origin from an excrescence and bending towards the wall of the intestine, of the hinder part of that œsophagus.

In regard to the structure of the remaining part of the intestine I do not wish to enter into any details. I only observed that the structure of the wall of that part which follows immediately after the œsophagus, and of the cœca, which penetrate as a rule as far as the sixth joint of the leg, is nearly the same. We find this wall everywhere beset with extremely numerous tubes or villi, which in some genera (*Nymphon*) are of a shorter and more rounded form, and in others (*Colossendeis*) are very slender and almost cylindrical. While the outermost part of the wall is formed of a single row of large distinct nucleated cells, these villi show a multi-cellular structure also. Each of these cells contains numerous globules, which for the most part seem to be of a fatty nature. The form of the cells is different, but they are commonly rounded. I call them cells, because each of the bodies has a distinct oval nucleus with a small nucleolus. As has been observed by almost all authors writing on the structure of the wall of the intestine, these cells often become detached from the wall, and are found lying loose in the contents of the alimentary canal. The nuclei in the cells of these villi were by no means easily observed in all the sections I studied. They were very distinct in the cells of the villi of *Colossendeis proboscidea*, Sab. (sp.).

It seems to be characteristic of the genus *Colossendeis* that the cœca destined for the different legs should branch off from the main duct, which runs straight from the proboscis to the abdomen, very close to one another, and close also to the place where the œsophagus communicates with the intestine ; at least I observed that in the three species of *Colossendeis* I studied (Pl. XVII. fig. 1). The number of lateral cœca given off at both sides of the main duct is six in *Colossendeis*. Of these the first two are rudimentary, one being the rudiment of those destined for the mandibles, the other (the first lateral pair) being that for the proboscis. Each of the four remaining cœca, of which the hindermost pass through a much longer part of the body before penetrating the leg than do the more anterior ones, shows a considerable swelling in the lateral processes, at the ends of which the legs are inserted.[1] The anal aperture of *Colossendeis* (Pl. XVII. fig. 1*d*) is an oval-shaped slit. It is not placed terminally or in the median line of the abdomen, but laterally.

[1] On a transverse section of the body of a *Colossendeis* between the lateral processes for the second and for the third pair of legs, five round sections of the intestine are seen placed close to one another. This furnishes a good opportunity of comparing their structure, but no difference is observed. Compare fig. 14 of Plate XXI.

In *Nymphon* the number of lateral cœca of the alimentary canal is five pairs. Of these the first pair is very wide and directed forwards. At the base of the proboscis it divides into two branches. One (the larger one) is directed upwards and forwards, and penetrates the mandibles; the other one enters the proboscis and divides, in some species (*Nymphon brachyrhynchus, e.g.*), again into two branches. These extend in some species farther than in others, but I never observed them beyond the hindermost half of its length.[1]

The groups of comparatively large cells with very thin walls and distinct nuclei, each of them containing, as a rule, one (sometimes more) strongly refracting granule probably have also some relation to nutritive functions. These I observed in *Nymphon, Phoxichilidium pilosum*, and in *Colossendeis*, collected in large groups sometimes about fifty in number. They seem not to be limited to any particular part of the body, but I found them always in the neighbourhood of the muscles, between the connective tissue, where fibrous threads keep them in place. I feel inclined to consider them as analogous to the fat-cells of most Arthropods. I figure a group of them in Pl. XXI. fig. 9.

5. *The Circulatory Apparatus.*—The somatic cavity is divided into distinct compartments, by means of sheets and bands of fibrous tissue. One of these, placed between the dorsal wall of the intestine and the dorsal integument of the body, is furnished with contractile walls, and has the function and the structure of a heart. In *Colossendeis* this heart is not surrounded by a pericardial sinus (Pl. XXI. fig. 14, *h*, Pl. XVII. fig. 1). The blood, entering the apertures of the heart, comes directly from one of the longitudinal compartments into which the somatic cavity is divided. The contractile walls of the heart do not enclose it on all sides; for on the dorsal side a part of the integument is used to form the dorsal wall of the heart. The contractility of these walls is due to the presence of muscles, which run in a transverse direction and are not striated. Along both sides of the heart these muscles are inserted into the dorsal integument of the body. As to their structure, I observed their fibres to be extremely thin and slender. When studying them with a strong lens (*e.g.*, 11, Immersion of Hartnack) I observed that they exhibit parallel edges only for a certain distance; for this parallelism almost imperceptibly passes over into an extremely feeble swelling of the fibre, in the interior of which a long nucleus with a distinct nucleolus is observed.

The heart of the Pycnogonids, as a rule,[2] is furnished with three pairs of apertures.

[1] In one specimen of *Nymphon brachyrhynchus* I observed that one of the branches penetrating the proboscis divided again, so that in the same section, through about the middle of the proboscis, five sections of intestinal cœca were observed. This I consider of no importance at all. It only proves, I believe, that it is almost dangerous to attribute any fundamental value to the number of pairs of cœca arising from the intestine. In a large specimen of *Pycnogonum litorale* a section of the fourth joint of the leg shows two sections of cœca in the same joint: the cœcum has given off a branch. Consequently I believe that the number of these branches depends in general upon the capacity of the different appendages. In *Nymphon* and *Colossendeis* no cœca are observed entering the palpi and the ovigerous legs, only because the capacity of these extremities does not allow of it.

[2] Not always. *Pallene brevirostris*, Johnston, *e.g.*, has only two of these.

The first pair are placed on the sides of the heart opposite to the second pair of legs; the second pair are placed similarly, but opposite to the third pair of legs; the third pair are found near one another at the posterior extremity of the heart. Through these three pairs the blood is admitted into the heart,[1] while it leaves it through a large opening placed at its anterior extremity. No aorta or arteries arise from it. In the heart of the specimens in spirits of *Nymphon robustum* and some other species I observed a compact mass of blood-plasm, which so totally filled up the cavity of the heart as to give, after having been taken out, an exact figure of its form.

As to the blood-corpuscles I only observed that they are, in *Nymphon* and *Colossendeis*, round and flat bodies with a distinct nucleus. However, I observed also more irregularly-shaped fusiform bodies, especially numerous in the cavities of the skin of *Colossendeis* (Pl. XVIII. fig. 1). With regard to their shape and dimensions there is no great conformity between the opinions of Cavanna and Dohrn; however, it is' only the study of fresh material that can finally settle such controversies.[2]

6. *Genital Organs.*—About the testis of the Pycnogonids hardly anything is known; and this cannot be wondered at when one considers that the true males were only discovered by Cavanna in the year 1875, the animals with swollen thighs described as males before that period being really the females. However, even Cavanna does not seem to have correctly identified the male organs, for he places them in the fourth joint of the legs. Dohrn has been the first, and hitherto the only one, who has pointed out the true position occupied by the testis, "Die Hoden liegen im Körper der Pycnogoniden, nicht in den Beinen, und bilden dort jederseits einen Schlauch, welcher in jede Extremität seiner Seite einen kleineren Schlauch absendet, der an der obenerwähnten Stelle in einer runden Oeffnung endet." This description, true in the main, is, however, not applicable to all the species of Pycnogonids; for, from what I have observed myself, I am able to furnish full evidence that, for some species, Dohrn's description is not quite correct.

A large specimen of a male *Colossendeis proboscidea*, figured of the natural size in Plate XXI. fig. 10, has been opened on the dorsal side. The skin with the heart being removed, the testis is observed *in situ*, and the intestine may be distinguished below it; the male organ, therefore, is placed rather at the dorsal side of the body.

The two laterally and longitudinally running parts of the organ are united posteriorly

[1] From observations I made, in the summer of 1880, in the laboratory of Prof. Lacaze-Duthiers at Roscoff.
[2] Cavanna calls them "piccolissimi globuli ellitici o sobellittici." Dohrn, on the other hand, describes the blood-corpuscles as "ungewöhnlich gross und complicirt." According to Dohrn, there are two forms of blood-corpuscles—"die einen bilden einen blassen, suzammengefalteten Ballon, in dem ein etwas glänzenderer linsenförmiger Kern sich findet neben 3-4 grossen Vacuolen; die anderen sind deutliche Amöben, mit lebhaften amöboiden Bewegungen und umschliessen oder tragen eine grössere Anzahl glänzender Tröpfchen." It seems to me that the large balloon-forming elements come very near to my fat cells (see p. 127); and as to the amoeboid corpuscles, these are probably my fusiform elements and the elliptic ones of Cavanna. If Dohrn is right about his large blood elements, then the animals must have still others of a third form.

by a commissure, so there is only one true testis, which has the form of a capital U. From the upstrokes of this U, on both sides, those parts originate which penetrate the leg, and which extend almost to the end of the fourth joint. The two parts of the testis which run longitudinally through the body are broad and flat; their course is not straight but rather undulated, being bent outwards whenever a lateral branch takes its origin. Although the organ is placed at the dorsal side of the body and legs, the male genital pores are situated ventrally towards the end of the second joint of each leg. This external opening is very small and is at the tip of a distinct conical tubercle. It leads into a canal which runs backwards almost parallel with the margin at the distal extremity of the joint and closely adheres to its chitinous covering; this canal runs along the wall of the joint till it reaches the dorsal side of the leg, here it becomes wider and turns inwards till it reaches the testis, which shows a small knob facing the beginning of the canal. Plate XXI. fig. 11, shows these particulars; the joint is figured as transparent, and the muscles which run from this to the following joint are to be considered as removed. The outer part of the canal, from the opening for about one-fourth of its length, is furnished with a thicker wall, and this shows most probably the extent to which the chitinous covering of the leg is bent inward. The remainder of the canal is formed of connective tissue,—at least I failed to observe an endothelial covering,— lined externally by a distinct muscular layer, the fibres of which run longitudinally and are not striated (Pl. XXI. fig. 13).

As seen in fig. 2, Plate XVIII., that part of the testis which penetrates the leg is in a transverse section kidney-shaped; a longitudinal duct is formed between the gland and the thin tissue or membrane which it touches laterally at two points. Most probably it is with the duct so formed that the canal which opens at the distal extremity of the second joint is in communication.

I have observed nearly the same structure of the male genital organs in the extremely transparent species of *Nymphon*, to which I have given the name of *Nymphon perlucidum* (p. 50 of this paper), and also in *Nymphon robustum*, Bell. Both species show genital pores in the male sex, only on the two hindmost legs; and quite in correspondence with this observation only two pairs of lateral excrescences of the U-shaped testis are to be seen. So when Dohrn says, for Pycnogonids in general, that the male organs penetrate each leg and open in a round pore on the ventral side (which is, no doubt, the case in all the species examined by him), he is laying down a rule which admits of a great many exceptions; for among the Pycnogonids dredged by H.M.S. Challenger there are species having their male genital pores only on the two hindmost legs; again, there are some which possess them on the three hindmost pair of legs, and, finally, there are species with pores on all the legs. My doubts about this point were cleared up by the observation of those species which bear their male genital pores at the tip of a stout cylindrical outgrowth (species of *Ammothea*, e.g.).

Of course the non-existence of a hardly distinguishable structure cannot be considered proved merely because the structure has not been observed, even in numerous specimens of the same species. But, on the other hand, if the structure is very easily seen in some specimens, we are at liberty to conclude that it is absent in those cases where it was not observed. Moreover, it is hardly possible that in those cases where the pores are only observed on the two hindmost legs, they should be present also on the foremost, and from some unknown circumstance should continually escape observation.[1]

In regard to the microscopical structure of the testis I have not much to say. On a transverse section numerous extremely minute cells are observed (see Pl. XXI. fig. 12), while that part of the testis which adjoins the longitudinal canal has lost its cellular structure, and shows a rather granular condition. Whether these granules are the spermatozoa, or whether the numerous globular bodies each furnished with a filamentary appendage on one side which I once observed, when pulling to pieces with needles a part of the testis, are the spermatozoa, cannot, of course, be ascertained from animals in alcohol, even when preserved so well as the Challenger specimens are. Only in mature animals do the male organs attain the development figured for *Colossendeis proboscidea*. This, most probably, is only the case during a short period of the year, considering moreover that the males in most species seem to be less numerous than the females (ten female *Colossendeis proboscidea* and only one male, twelve female *Colossendeis leptorhynchus*, and only one male, &c.), it cannot be wondered at that the number of species in which I could investigate these organs was limited. Except in *Colossendeis proboscidea*, and in three species of *Nymphon*, I observed the testis also in the leg of *Ascorhynchus glaber* (Pl. XVI. fig. 9, *n*), but here only in the fourth joint of the leg.

In the other specimens which I consider as males, the only means I had to make out the sex consisted in looking for external sexual characteristics, such as are afforded by the dimensions of the genital pores and the condition of the thighs. Large genital pores and swollen thighs are characteristic of the females; slender thighs and small pores, very often not present in the first or in the first two pairs of legs, are characteristic of the males. Moreover, a transverse section of the thigh of one of the legs is easily made, and does no injury worth mentioning to the specimen. When in such a section no ovary is observed, so far as my experience goes, it is almost certain that the

---

[1] The genital pores of *Nymphon robustum*, Bell, ♂, are tolerably large, and are easily observed with the aid of a magnifying glass. Yet I have examined large specimens of this species (dredged in the Barents Sea), where these pores were not observed, even when investigating the joints with the microscope. As I was convinced of the exactness of this observation, I felt greatly puzzled with it at first; afterwards on reading a paper of Schöbl in the Archiv f. Mikroskop. Anatomie, Bd. xvii., 1860 (Ueber die Fortpflanzung isopoder Crustaceen) I found that this author admits that in the females of these Crustaceans, the genital pores are only present at a certain period, and are totally wanting during the rest of the year. Perhaps there are male Pycnogonids which have the same peculiarity.

specimen is a male ; even in very young females at least a trace of the ovary is always to be observed in those joints.

According to Dohrn the female glands are disposed in the same way as the testis, with the only difference, that in the ovaries of most species, the lateral excrescences which penetrate the legs and often force their way into the penultimate joint, are the only parts which develop mature eggs.

What I happened to observe agrees almost completely with Dohrn's description. I perfectly believe, however, that the occurrence of a part of the ovary in the body of a Pycnogonid is rather rare, because, as a rule, only the lateral excrescences remain. No doubt this must be considered as a secondary condition; and seeing that, so far as I could ascertain, it is the rule in all the species of *Nymphon*, my original opinion, that the genus *Nymphon*, of all the genera of Pycnogonida, resembled most the hypothetical ancestors of our group, was severely shaken.

The genus *Colossendeis* affords a beautiful example of the original condition of the ovary. I examined a specimen of *Colossendeis leptorhynchus*, Hoek. A transverse section of the body showed the place occupied by the ovary, and convinced me of the exactness of Dohrn's assertion. However, as in the case of the testis, Dohrn's observation is incomplete in one respect ; there are not two ovaries, but only one. Immediately in front of the abdomen the two lateral parts are united by a cross-piece, and so the ovary has the very same shape as the testis, viz., that of a U with long upstrokes. That it occupies also the same position in the body as is the case with the testis, is clearly seen in fig. 14, Plate XXI., which is a drawing of a transverse section of the body. In *o.* the ovary is shown placed above the intestinal tract and its excrescences which penetrate the legs. The dimensions of the eggs in that part of the ovary which is situated in the body are, at least in this species, the same as of those eggs which are found in the thighs of the legs. The lateral excrescences penetrate all the legs, and in the specimen I studied reach the sixth joint. This, most probably, will greatly depend on the state of maturity of the animal. I often found the eggs only in the thigh, but often also (*Hannonia typica*, Hoek, Pl. XIV. fig. 11, *e.g.*) the eggs are found as far as the end of the second tibial joint. The ovary is always placed dorsally to the intestinal tract, which is also the case with those animals which have only the lateral excrescences left, and thus show several ovaries.

The structure and the formation of the ovarian eggs I will describe further on. In regard to the way in which the eggs are laid, I had the good fortune to observe the copulation of a male and female *Phoxichilus laevis*, Grube, when I was, last summer, in the zoological station of Professor H. de Lacaze-Duthiers at Roscoff. The eggs are fecundated the moment they are laid, and the copulation, therefore, is quite external, brought about by the genital openings of the two sexes being placed against each other. Half an hour after the beginning of copulation, the male had a large white egg-mass on one of his ovigerous legs, and about one hour later both masses were present.

The female genital openings.are a great deal larger than those of the male, and are of an ovate shape, and, as a rule, oviducts are totally wanting. I only observed them in the genus *Colossendeis*, where they have nearly the same course as the vasa deferentia. Nearly in the middle of the second coxal joint of the two hind legs a lateral branch arises from the ovarian cœcum, which passes through the joint. The interior of this branch, which is the oviduct, is in immediate communication with the ovary, and during the breeding season the eggs are found penetrating the branch. While the ovarian cœca which penetrate the legs are lined only by connective tissue, the oviducts which begin at the dorsal side of the second coxal joint and run along the wall of the joint till they reach the opening on the ventral side, are lined by a coat of longitudinal muscle fibres. The female genital opening in this species is small and rather triangular.[1]

From the end of the oviduct which reaches the opening, distinct muscle fibres radiate, and are inserted round the opening on the inside of the chitinous wall of the joint. A transverse section of the fourth joint of the leg of a female *Colossendeis leptorhynchus* is figured in fig. 16, Plate XVI. ; *m* is the ovarian cœcum which, as this specimen is by no means mature, is only of small dimensions ; when in the breeding season, the whole central cavity (which in the figure is represented as empty) is filled with eggs ; it swells to such an extent as to fill up almost the whole cavity of the leg, in so far as this is not occupied by the intestinal cœcum. As in the other species of *Colossendeis*, the eggs are extremely numerous and small. Each ovarian egg has a distinct germinal vesicle, which is placed almost exactly in the centre of the egg, and has, as a rule, one distinct and very glittering germinal spot. Among the older eggs, which are richly furnished with yelk, smaller ones are always observed whose protoplasm is almost quite transparent. Each egg is surrounded by a very thin membrane, which is a true "vitelline membrane" and adheres closely to the protoplasm of the egg.

In the genus *Nymphon*, I have investigated the female genital organs of the following species :—*Nymphon brevicaudatum*, Miers ; *N. brachyrhynchus*, Hoek ; *N. robustum*, Bell ; *N. longicoxa*, Hoek ; and *N. hamatum*, Hoek. I never observed the ovary in the body of these species, and always found at least a trace of it in the fourth joint of the leg (thigh).

When in an early stage.of development, the ovary is placed against and at the dorsal side of the intestinal cœcum which passes through the joint. While the ripe ova, which are often very large (Pl. XVI. fig. 7 *l*, ripe ovum of *Nymphon brevicaudatum*, Miers), have a thin vitelline membrane as in the ovum of *Colossendeis*; younger ova have often the

---

[1] In this species, as I have mentioned already (p. 63), the genital openings, both in males and females, are present only on the two hind legs ; at least in all the specimens at my disposition, I failed to observe them on the first two pairs of legs. However, ovarian cœca penetrate the first two pairs of legs as well as the two others.

appearance of being surrounded by a very thick and concentrically stratified membrane (ovarian egg of *Nymphon brevicaudatum*, Pl. XXI. fig. 15). In other species (*Nymphon robustum*, Pl. XXI. fig. 16) this membrane is perfectly transparent, while in both cases the granular protoplasm of the egg within this thick membrane or capsule seems to be surrounded by its vitelline membrane. In a third case, which I observed in *Nymphon longicoxa* (Pl. XXI. fig. 17), the membrane of the egg does not seem to be thick, but irregularly folded and crumpled; yelk-particles are here very numerous in the central part of the egg, which surrounds the germinal vesicle; and the protoplasm of the egg extends beyond this central part till it reaches the crumpled membrane. However, it is very probable that the condition of the eggs has suffered from their having been so long in alcohol, only I wish to point out that from what I observed it is almost certain that any one studying the formation and the development of the animal egg, will find a very interesting object in the egg of the Pycnogonida. The dimensions of the mature ova are very different. Of the specimens I studied they are largest in *Nymphon longicoxa*, *N. brevicaudatum*, and *N. robustum*, a great deal smaller in *N. brachyrhynchus* and *N. hamatum* (the number of eggs united in an egg-mass being always in inverse proportion to their size).

While in the younger ovarian eggs the germinal vesicles as a rule are placed in the centre of the egg, in the very large mature egg the vesicle is placed close to the wall. Sometimes (*Nymphon longicoxa*) it has the shape of a sand-glass, and once I observed an extremely small micropyle canal in the membrane of the egg, just opposite the place occupied by the germinal vesicle. As a rule there is only one germinal spot, but I once observed two distinct spots in the germinal vesicle of the egg of *Nymphon longicoxa*. In the ovarian egg of *Nymphon robustum* one distinct nucleolus may be observed almost exactly in the centre of the rounded and granular germinal spot. As for the manner in which the eggs make their way to the genital apertures in those cases in which no true oviduct is observed, I think there can be no doubt that the body-cavity itself performs the function of an oviduct. The absence of such a duct at the genital pores, and the fact that I repeatedly observed detached eggs pressed against the connective tissue surrounding the ventral ganglia, or other parts in the interior of the body admits of no doubt in this respect.

All I have said about the ovary and the formation of the ova in the genus *Nymphon* also holds good in the case of the other genera. The limited quantity of specimens prevented me from making a section of the body of species of these genera. Most probably *Ascorchynchus* will show the same disposition as *Colossendeis*. A transverse section of the thigh of one of the legs of *Ascorchynchus orthorhynchus* is figured in Plate XVI. fig. 11. The thigh is much more dilated than one of the other joints of the leg, yet it is not round but flattened, and the contents are almost divided into two unequal parts by the large chitinous thickening, which at the one side is in connection with the wall of the leg. The one part is filled up with the ovary, the other partly with the cœcum of the intestine. The ovarian eggs are small and are furnished with a central germinal vesicle. Most probably the females of the species of *Ascorhynchus* have an oviduct like that of the species of *Colossendeis*. The species of *Pallene* show the same disposition as in *Nymphon*. A transverse section of the thigh of *Pallene australiensis* corresponds perfectly with that of *Nymphon brevicaudatum*, figured in Plate XVI. fig. 7 ; there is one very large and probably mature egg with an eccentric germinal vesicle, and numerous smaller ovarian eggs, with their vesicle in the centre. Moreover, minute researches on *Pallene brevirostris* (an inhabitant of the Dutch coast) admit of no doubt as to the structure of the ovaries ; they are totally wanting in the body, and take their origin in the thighs of the legs. The eggs when mature are large, and their number is limited.

In the genus *Phoxichilidium* I studied the anatomy of the body of *Phoxichilidium pilosum* (a female specimen) without meeting with the ovary. In the legs of this species, however, and also of *Phoxichilidium patagonicum*, I soon found it. The whole cavity of the leg is often filled up with eggs, and these are even observed pressed closely against the wall of the leg (Pl. XVI. fig. 17). The eggs are comparatively small and very numerous. The membrane of the eggs is much thicker than is the case with the eggs of the other genera (Pl. XXI. fig. 18). Neither in the species of *Pallene* nor of *Phoxichilidium* did I observe the least trace of an oviduct, so I think that here, as in *Nymphon*, the genital aperture communicates directly with the cavity of the leg. I think also that the circumstance I often observed of eggs free in the cavity of the leg is in support of this opinion (Pl. XXI. fig. 18). For the other genera of Pycnogonids I have, so far as the ovaries are concerned, no observations worth mentioning.

7. *Observations on the Embryology of the Pycnogonida.*—Among the Pycnogonids of the Challenger Expedition there were some species provided with eggs. On account of the great importance of embryology for the study of the affinities of a group of animals I tried to acquire as much information on this subject as possible. Unfortunately, with the exception of one species belonging to the genus *Ascorhynchus*, G. O. Sars, all the species with egg-masses belong to the genus *Nymphon*, Fabr., but of these there are out of twelve species no less than six provided with eggs.

The researches of Kröyer, Johnston, Goodsir, Dohrn, Semper, Cavanna, and myself, have shown that in the genera *Pycnogonum*, *Pallene*, *Phoxichilus*, *Phoxichilidium*, *Nymphon*, &c., the eggs after having been laid are carried on the so-called ovigerous legs. The honour of having discovered that not the females (as was believed by the older authors) but the males fulfil the duty of bearing these eggs is due to Cavanna ; this observation has since been confirmed by the researches of Dohrn, Böhm, and myself.[1]

---

[1] The observations of Cavanna were published in the year 1875. It is indeed strange to see that neither Wilson nor Miers have heard of this discovery. These authors, in their descriptions of new species, &c., are therefore almost constantly confounding the two sexes.

In the first place, however, I wish to draw special attention to the fact that with regard to *Nymphon brevicaudatum*, Miers, this rule admits of an exception. I examined a species with large genital pores and swollen thighs, and provided with egg-masses on the ovigerous legs. On investigating transverse sections of the thighs, I soon saw that this specimen was a female. So far as I know, this is the first time that an exception to this rule has been observed. In the second place, I wish in a few words to discuss the circumstance that, although eight different species of the genus *Colossendeis* were collected (together represented by thirty-one specimens, and four specimens of *Colossendeis proboscidea*, Sab. (sp.), trawled north of Scotland during the cruise of the "Knight Errant"), none of these are provided with eggs. The number of males, however, is very restricted : there is only one male *Colossendeis leptorhynchus* among nine specimens, there is one male *Colossendeis gigas* among six specimens, one male *Colossendeis megalonyx* among seven, and, finally, one male *Colossendeis brevipes*. On the other hand it is possible that the genus *Colossendeis* is an exception to the rule, and that the males in this genus may not have the gallantry to nurse their babies as the males of the species of other genera are accustomed to do. For, comparing the ovigerous legs of the males with those of the females, a distinct difference is almost always easily observed : those of the males are a great deal stouter, the fifth joint is as a rule swollen towards the extremity, or furnished with a distinct knob, &c. ; but in the ovigerous legs of the males of the species of *Colossendeis*, these differences in form and size are never observed. So it is quite possible that they deal differently with their eggs from the species of other genera.

The species provided with eggs are: *Nymphon hamatum*, Hoek ; *N. longicoxa*, Hoek ; *N. fuscum*, Hoek ; *N. brevicollum*, Hoek ; *N. brachyrhynchus*, Hoek ; *N. brevicaudatum*, Miers ; and *Ascorhynchus minutus*, Hoek. Of the latter species there are in all two specimens, and of these one bears eggs. But the development of these eggs is in its last stage, so that I was only able to ascertain the form of the larvæ. The eggs of this species are extremely small, and at the same time numerous.

It consequently happened that my embryological researches were limited to the genus *Nymphon*; in so far not unfavourable, as yet almost nothing has been published on the embryology of this genus.

Full-grown males of the genus *Nymphon* bear the eggs on the fourth and fifth joints of the ovigerous leg, or only on the fifth joint ; the curious foliaceous appendages occur on the sixth to the tenth joints of the leg, and have nothing to do with the egg-bearing function of the leg. Yet it is possible that they may be of some use in seizing the eggs when just laid, but, on the other hand it must be observed, that in the genera where these appendages occur, the ovigerous legs of the females are furnished with them as well as those of the males.

The eggs are soldered together and form in the species of *Nymphon* I studied, and in *Ascorhynchus minutus*, one packet only on each ovigerous leg. I believe, however, that cases are by no means rare in which two or even three packets are formed on one ovigerous leg. The packet is placed round the leg, and in some species (*Nymphon brachyrhynchus*, Hoek, e.g., Pl. XIX. fig. 1), it may be easily brushed off. In other species, however, *Nymphon robustum*, Bell, for example, this is by no means so easy, the surface of the joints of the ovigerous legs being furnished with numerous hook-like spines.

Every egg in the packet has its own membrane, a very thin and structureless tunic. The size of the packets is very different. It varies greatly with the size of the animal, but is different also in different specimens of the same species. The size of the egg seems to be constant for every species ; consequently the size of the packet will depend on the number of eggs in each packet ; and the age and the condition of the female will, of course, influence this.

The egg of *Nymphon brevicaudatum*, Miers, measures 0·5 or 0·7 mm. ; when in the first stages of development it is nearly globular (0·6 × 0·6 mm.), afterwards oval (0·5 × 0·7 mm.). The number of eggs in each packet varies between fifty and sixty. The egg of *Nymphon fuscum* is a great deal smaller (0·15 or 0·12 mm. The eggs of *Nymphon brachyrhynchus* are about 0·55 mm. and even in a much advanced stage of development nearly globular. Large packets of the latter species do not contain more than fifty eggs, and the dimensions of these packets vary between 3·14 × 1·85 mm. and 2·3 × 1·6 mm. The egg of *Nymphon brevicollum* has a diameter of 0·26 mm.; the number of eggs in a packet is in this species about ninety. Fig. 2 on Plate XIX. gives a section of a packet of eggs of this species. The colours are those which are seen when the object is coloured with picrocarmine, and the figure is half in outline ; every egg is seen to be placed in a cavity formed by the cement which solders the eggs together (*c*), and coloured distinctly red by the picrocarmine. The large opening (*o*) in the centre is that occupied by the ovigerous leg ; the small holes (*s*, *s*) are those which are left between the eggs when soldered together. On the outside mud and sand particles adhere to the packet (*m*).

So far as I know, Dohrn is the only author who has published observations on the cleavage of the Pycnogonid egg ;[1] but as the method of making sections of such very small eggs was not yet in use when he published his paper, and could not, therefore, be applied by him, I might reasonably have expected to see much more than he did, by availing myself of this method of recent embryology. Yet my researches in this respect were not very successful, owing at least partly, I believe, to the condition of the material I studied. Every one will acknowledge how necessary it is, especially in embryological researches, to study fresh and also very rich material ; now the Challenger Pycnogonids had been six or seven years in alcohol before I studied their eggs, and, moreover, the

---

[1] A. Dohrn, Ueber Entwicklung und Bau der Pycnogoniden, Jenaische Zeitschrift, Bd. v., 1869.

quantity was limited, so I need not appeal to the indulgence of the reader on account of the imperfection of my researches in this department.

The study of the eggs of *Nymphon brevicaudatum*, Miers, was the most successful of all. These eggs are the largest of the species here in question ; the number of animals furnished with eggs was in this species rather great ;[1] and their condition was superior to that of the eggs of the other species. The method I followed is well known. I enclosed the eggs (hardened with absolute alcohol) in paraffine, and coloured the sections afterwards with picrocarmine.

Fig. 3 is a drawing of the first stage I was able to observe. The food-yolk and the formative-yolk (deuto- and proto-plasma, Ed. van Beneden) are still mixed together, and the cleavage is complete. Every segment is furnished with a nucleus, coloured distinctly red by the picrocarmine, and situated almost in the middle of each segment. The structure of the yolk particles in each segment is very curious, and probably this is caused by the continued action of the alcohol. In fig. 4 I give a strongly-magnified drawing of a small part of such a segment just at the border of the section. It looks as if the yolk-elements had grown vesicular,—a matter I only make mention of as the same structure is no longer observed in the next stage of development of the egg. In this stage, as in the following, the egg is furnished with a distinct but very thin membrane.[2]

The second stage I observed has the blastoderm distinctly developed. The cells of which it is composed are very much flattened, and do not show distinct limits ; a very large nucleus is, on the contrary, always easily observed. Fig. 5 shows the cells as seen on section, fig. 6, the blastoderm with the nuclei magnified. Every nucleus shows a distinct nucleolus and numerous small granules. In this stage the food-yolk is irregularly split into larger or smaller parts, which are coloured yellow by the picrocarmine ; they do not show the vesicular structure of the yolk-segments in the first stage, and are not furnished with a nucleus.

A transverse section of the next stage of development I observed is figured in fig. 7.[3] Here the embryonic development is already far advanced, consequently I was not able

---

[1] The eggs of *Nymphon hamatum*, *N. longicoxa*, and *N. fuscum* were so far advanced in development that in them only the different larval stages could be studied.

[2] Dohrn, *loc. cit.*, p. 139, says the egg of *Pycnogonum litorale* has a double membrane, and that these membranes are found in the ovary, an assertion not corresponding with the observations I made on the eggs of *Nymphon*.

[3] Between the stage figured in fig. 7 and the foregoing, numerous other stages were observed ; but in these the cellular structure was so totally spoiled by the action of the alcohol, that I dare not give drawings or descriptions of them. The only means of distinguishing the embryonic cells from the deutoplasm is by the colouring of the cells with picrocarmine, and there can be little doubt that one of the first changes the blastoderm undergoes consists in the formation of a longitudinal thickening of it at the future ventral side of the embryo. This thickening terminates rather abruptly at the anterior end, but at the posterior end it slopes gradually to the unicellular part of the blastoderm. Afterwards a longitudinal furrow seems to take rise in the middle of this thickening, the inner part of which is finally isolated in the form of a longitudinal tube. I publish these details only with the strongest reserve, the condition of the eggs and the circumstance that the sections are necessarily taken in quite uncertain directions, making the giving of a decided description impossible.

to study the formation of the germ-layers, nor the modifications which they undergo during development. Whether all the cells of the embryo in this stage are derived exclusively from the blastoderm, or whether they are also partly due to the deutoplasm is a question which it is impossible to answer from the section before me. Dorsally the greater part of the embryo is covered by a single row of flattened cells (the original blastoderm cells), ventrally a plate is clearly distinguished much thicker than the blasto-derm, and doubtless formed of cells more than one row deep. Unfortunately, however, the limits of these cells were quite gone ; I therefore could not distinguish either their number or arrangement, but I believe the evidence is great that in the inner layer of this plate the original mesoblast is to be seen. In this stage rudiments of the appendages are distinctly formed ; and I consider it a very characteristic feature in the development of the Pycnogonids, that the food-yolk penetrates into these appendages. In the section here figured, however, that part of the food-yolk which penetrates the leg, is not in direct connection with the central food-yolk mass ; but this is caused by the circumstance, that the section does not pass exactly vertically through the embryo, but goes a little obliquely from above backwards to the ventral side.

The blastoderm shows to a considerable extent in the stage I have figured a double cell-layer dorsally in the middle, and even a small lumen is observed between these two. Small cells or nuclei seem to be present in this lumen, and the whole arrangement made me think it possible that I had an early stage in the development of the heart before me. The broad and flattened condition of the heart in the adult animal of *Nymphon* is not opposed to this suggestion ; yet it is difficult to understand why a heart should be developed before there seems to be any question of an intestinal tract.

About the same stage is also figured in figs. 9 and 10. At the ventral side the first pair of appendages (the foot-jaws), three pairs of legs, between the foot-jaws the proboscis, and the caudal protuberance, are easily distinguished. The second and third pair of cephalic appendages show in this species a remarkable retardation in their appearance, visible in the stage in which the first and second pair of true legs are already two-jointed and bent inwards so as to meet in the middle of the ventral surface, and in which the third pair is longer, yet bent inwards and forwards. In this same stage the third cephalic appendage is not yet distinguishable, and the second pair only shows a small protuberance at the base of the foot-jaws. An equatorial section of an embryo in this same stage is figured in fig. 11. Between the foot-jaws (*a*) and the first true leg (*b*) two small protuberances are distinguished, the first of which (*c*) is larger than the second (*d*), which in this stage is observed only interiorly. The section is also remarkable for the distinctness with which the nerve ganglia are seen.

There is good reason to consider this arrangement characteristic for the species *Nymphon brevicaudatum*, Miers. Other species of *Nymphon*, of course, may show the same ; so far as I could ascertain it is not the rule, for neither *Nymphon brachyrhynchus* nor *N. brevicollum* nor *N. hamatum* agree with *N. brevicaudatum* in this regard. Fig. 8 shows a stage in the development of *Nymphon brachyrhynchus*, in which the three first pairs of embryonic appendages are already present. The first pair (the largest) are armed with pincers; the second and third are small, armed with curved hooks and not taking parts of the food-yolk; of the true legs in this stage nothing as yet is to be seen. In figs. 12 and 13 I have figured a larva of *Nymphon brevicollum* showing the three cephalic appendages, the first pair of true legs almost completely developed, the second pair much shorter than the first, and not yet furnished with claws, the third only as a rudimentary process; the fourth pair is totally wanting in this stage. Consequently I believe it is the rule in *Nymphon*, that the three pair of cephalic appendages are developed first of all, the legs appearing afterwards in regular succession.

To return to *Nymphon brevicaudatum*, Miers, in figs. 9 and 10, I have figured embryos within the shell of the egg almost of the same stage; with this difference only, that in fig. 10 the egg is figured as seen from the ventral surface. In fig. 9 also, a part of the dorsal surface being bent over to the ventral side has been drawn. In this last figure it is clearly shown that the dorsal surface of the embryo is at least at the anterior side lined with a shell-like thickening, the proboscis and the first pair of cephalic appendages being at their origin covered by this thickening as by a cap. Near the anterior side of this cap the double supra-œsophageal ganglion is situated, making it evident that in the border of this cap the anterior margin of the cephalic part of the embryo is to be seen; the proboscis being only an azygous excrescence of that part of the ventral surface which surrounds the mouth.

The equatorial section figured in fig. 11 shows the distribution of the nerve ganglia on the ventral surface; the first and second ganglia are smaller and are placed close to each other; the development of the third, fourth, and fifth ganglia is in near relation with that of the corresponding legs; finally, neither the sixth ganglion nor the fourth pair of legs is to be distinguished. In the middle the two halves of every ganglion are placed close to each other, which, as far as I could ascertain, is also the case in earlier stages. Of the longitudinal commissures between the ganglia in this stage, nothing as yet is to be distinguished, and as to the cellular structure of the ganglia, I was only able to trace large cells without any differentiation.

The degree of development the larvæ have reached when leaving the shell of the egg is not the same for all the species of *Nymphon*; so I think it probable that the larva of *Nymphon brevicaudatum*, Miers, does not creep out of the egg before the four true legs are developed, whereas the young of *Nymphon brevicollum* cling to the ovigerous legs of the father as soon as only one of the pairs of true legs has reached its full development, and perhaps even earlier yet. So, when Semper affirms that there occurs a complete metamorphosis in the development of the species of the genus *Nymphon*, two points are to be borne in mind, (1) that this does not affect all the species of *Nymphon* in the same way, and (2) that here the word metamorphosis has quite a different meaning from what it has in entomology.

Of the genus *Nymphon* I was able to compare the larvæ of the species *Nymphon brevicollum*, *N. hamatum*, and *N. longicoxa*. Of *Nymphon brevicollum* I have figured the youngest stage observed in figs. 12 and 13; an older one, which has three pairs of legs fully developed and the fourth already planned in the form of two lateral processes, has been drawn from the ventral side in fig. 1 of Plate XX.

On the ovigerous legs of the same animal I found together larvæ in both the stages I have figured, and also in intermediate stages. Taking a small number of these larvæ from the leg to study them under the microscope, I often observed the membranes of earlier stages between them. These membranes, and especially the parts which belong to the fore-part of the body, are attached to one another by means of long threads; these threads take their origin in the first joint of the foot-jaw, which bears a protuberance perforated by the thread. In the interior of the joint, and also of the empty membrane of this joint the thread can be traced a short way, but in neither to a great extent, as in the joint it is covered by the food-yolk, and in the membrane soon ceases after having passed the protuberance.

The larvæ of *Nymphon hamatum* which I was able to study were already furnished with four legs. Their condition was not extremely favourable for minute investigation, especially because the food-yolk makes the whole body opaque. The third pair of cephalic appendages are but small, and have each the form of a two-jointed stump bearing a pair of small spines at the extremity. The fore-part of the body of this larva is figured in fig. 3, Plate XX. An apparatus of a very singular shape, and, of course, closely connected with the protuberance perforated by the long thread in the larva of *Nymphon brevicollum*, is situated as in that species in the first joint of the foot-jaw. Numerous bottle-shaped sacs are placed near each other, and in such a way that their necks meet in one point. Each neck terminates in a small semilunar border, which covers a small slit; through this slit a thread passes, that can be easily observed as it runs through the throat of the bottle-shaped sac. The widened part of the bottle has in its interior two or more vesicles, which seem to be filled with an opaque protoplasm, covering in all probability the origin of the thread. Every bottle has its own thread, and of these more than ten are easily counted. I have figured this apparatus in fig. 4, Plate XX. The study of the apparatus is very difficult, as it is not transparent, being covered at one side by the food-yolk. The different bottle-shaped sacs are enclosed in a granular mass, with which very fine fibres seem to correspond. I could follow these fibres to a certain distance from the apparatus, where they are covered by the food-yolk; and from their pale appearance, and the circumstance that they are not easily coloured by picrocarmine (as the muscles, fig. 4, are), I felt inclined to look upon them as nerve-fibres.

The same organ, but of a somewhat different shape, occurs also in the mandibles of the larvæ of *Nymphon longicoxa*. These larvæ in almost every detail correspond with those of *Nymphon hamatum*, but the organ here in question is a curious exception. I have figured it in fig. 5 of Plate XX. Numerous small vesicles are in close relation with each other, and are so placed that they seem to radiate from a common centre. The whole apparatus is small; fig. 5 shows a drawing enlarged 270 times. The vesicle corresponds in all probability with a larger one placed in the centre, sending forth the thread, which in this species is always a single one. The thread is a little swollen at the foot, and seems (to judge from the double lining under the skin) to run through a sheath, at the end of which a semilunar border covers a small slit, through which the thread passes. Of course the chitinous skin of the larva is not coloured by the picrocarmine, but the broad and flat thread is. This thread is very long, its length sometimes equalling and even surpassing half the length of the larva; at a certain distance from the beginning I repeatedly observed a small part of the old skin, which remained in relation with the thread, while the larva got a new one.

A similar apparatus to the one described by me for the larvæ of *Nymphon hamatum* and *Nymphon longicoxa* (and occurring in all probability also in *Nymphon brevicollum*) has been observed by Dohrn in the larva of *Achelia*. The first joint of the mandible bears a strong spine in the larva. "An diesem letzteren sieht man fast immer einen sehr feinen Faden befestigt und erkennt bei näherer Untersuchung, dass dieser Faden aus dem Dorn herauskommt. Der Dorn ist nämlich hohl, seine Spitze durchbohrt und im Inneren sieht man einen zweiten feinen Canal der von einem merkwürdig gestalteten Organ ausgeht, das in der Basis des ersten Gliedes der Scheerenfüsse liegt. Das Organ hat die Gestalt eines Kartenherzens, die Spitze ist verlängert in den eben erwähnten Canal, der anfänglich etwas breiter sich bald verschmälert und quer durch den Innenraum des Beines sich zu dem Dorn begiebt. Der Canal ist nicht häutig, sondern hornig, dennoch beugt er sich in mässiger Krümmung, ehe er den Dorn erreicht. Die Structur der Drüse—denn für eine solche muss ich das sonderbare Organ halten—habe ich nicht ermitteln können, nur so viel vermag ich anzugeben, dass die hintere Hälfte aus kleinen Zellen bestand, die dem Organ eine gewisse Aehnlichkeit mit einem Nervenganglion verliehen, während die vördere Hälfte von zwei merkwürdigen blassen Flecken eingenommen wurde, die Kugelgestalt besitzen, aber nicht erkennen liessen, ob sie mit irgend einer Substanz gefüllt waren, oder Hohlkugeln darstellten. Ueber und unter dieser Drüse liegen Muskeln, welche zur Bewegung des zweiten Gliedes der Extremität dienen."[1]

There remains no doubt that the organ of Dohrn is the same as that observed by me. Also, as I already said above, I feel very much inclined to adopt the conjecture about the character of the organ proposed by him. The organ is a gland, and the product of its secretion consists of one or more fine threads. These threads occur only in the larval condition, and as for their use I wish to compare them with the byssus threads of the Lamelli-

[1] Dohrn, *l. c.*, p. 141.

branchiata. The larvæ of different species, as observed by me, usually remain for a long time after having cast off their exuviæ, in relation to the ovigerous leg of their parent. As long as they were enclosed in their egg, they clung together tightly enough ; but once crept out of the egg-shell, a special arrangement is necessary to keep them together. This is found in the threads, and the supposed glands from which these take their origin, as observed by Dohrn and me. Repeatedly I saw, as I have mentioned already before, between the larvæ of *Nymphon brevicollum*, collections of very numerous skins held together by means of the threads, and small parts of such a cast skin I found also in relation with the thread of the larvæ of *Nymphon longicoxa*.

It is true that the structure of the apparatus, as it shows itself in the larva of *Nymphon hamatum*, argues, perhaps, for the conjecture that the organ is an organ of sense, but then it is exceedingly strange that such an organ should only be found within the larvæ. And it would be difficult to explain the meaning of the single or numerous long threads as being sent forth from an organ of sense, whereas in relation with a gland their function can easily be understood.

The study of this same organ which I made last summer in the laboratory of Prof. Lacaze-Duthiers, at Roscoff, has also convinced me, that my original supposition as to the function of these organs was erroneous. The fine threads, which I observed in the interior of the mandible running towards the organ are threads of connective tissue ; their function is, no doubt, to hold the organ in its place. The young of *Nymphon robustum*, Bell, and those of *Nymphon brevicaudatum*, Miers, are a great deal more developed when creeping out of the egg than those of *Nymphon hamatum*, *N. longicoxa*, and *N. brevicollum*. Most probably this spinneret of the larva does not occur in these species.

Besides the larvæ of the genus *Nymphon*, the only other genus of which I could investigate the larvæ was *Ascorhynchus*. About their development and metamorphosis nothing as yet has been published. I can only give a drawing of the single larval stage which I observed, and which is furnished with three pairs of legs. The fig. 6 on Plate XX. shows that the larva in this stage corresponds with larvæ of other genera, as observed by Kröyer, Dohrn, and myself. Of the glands in the foot-jaw no trace could be discovered; but then the larvæ are very small, and their condition is not very good.

8. In studying the anatomy of the Pycnogonids of the Challenger Expedition, I met with two different kinds of bodies of which I have not been able to ascertain whether they really belong to the organisation of the Pycnogonids, or must be considered as parasites. However, I feel much inclined to adopt the latter opinion ; and although some doubt remains, I wish to give a short description of what I have seen, which may, perhaps, be of use for later investigators.

In the first place I met with some curiously shaped forms in the interior of the body and of the legs of two different species of *Nymphon*. I observed them in *Nymphon longicoxa* and in *Nymphon brevicaudatum*, but only in some of the specimens which belonged to the male as well as to the female sex. They are comparatively large, often long ovate cells with a thin wall, the contents consisting of large granules and a longitudinal slightly curved nucleus.

When colouring the preparations with picrocarmine these forms assume a yellow colour, the nucleus becoming beautifully red. They seem to be distributed through the body very irregularly and seem to penetrate all the cavities accessible to the blood. In *Nymphon brevicaudatum* I even observed them in the space before occupied by the eggs, and in which still an unripe egg was to be seen. Both in this species and in *Nymphon longicoxa* most of these curious forms are very regularly placed against the wall of the leg, where they often form two or even more distinct layers. Their size varies between 0·066 and 0·081 mm. With regard to their nature my opinion is not at all a settled one; but I am strongly inclined to believe them to be the eggs of some parasitic animal. But what kind of animal their parent in that case will prove to be I am unable to say.

The other kind of bodies must be regarded, I believe, as ectoparasites of *Colossendeis leptorhynchus*. Of the ten specimens of this species in the Challenger collection there are three which are sprinkled over with these. The one is a male, the two others are female.

They are rounded, sac-forming bodies, often with a crumpled surface placed at the end of a short stalk, the end of the stalk is in connection with the integument of the Pycnogonid. Their wall is chitinous, and under this outer wall there is a much thinner inner one; in the stalk this inner wall is close to the outer one, but in the globular part there is a large open space between the outer and the much smaller inner sac. In this space pressed against the outer sac numerous eggs are found, the size of which is 0·088 mm., they have a very thin wall and are furnished with a yelk of large rounded elements, coloured yellow by picrocarmine. A small nucleus as a red coloured spot, however, is always present.

Whether these are really eggs is the first question to be answered, and I think there can be no doubt in respect to this. Moreover, to judge from their structure and that of the capsules, they are eggs that are laid after having been fecundated. There is only one consideration, I believe, that may be set against this suggestion, and this is, that all these eggs are in the same state of development; not only those of the same capsule, but of all the capsules I investigated. It must be borne in mind, however, that these were brought up by the same haul of the trawl, and probably lived in the neighbourhood of one another; consequently I think this objection is of no importance.

The second question is whether they are the eggs of that Pycnogonid on the legs of which they are found, or of another specimen of the same species, or of any other animal. Of course it is possible that the eggs are *Colossendeis* eggs; however, I do not think this very probable. In the first place, because males and females both are studded with these capsules, and in the second place, because these capsules are totally different from the egg-masses commonly found on the ovigerous legs of the Pycnogonids. In favour of

144 THE VOYAGE OF H.M.S. CHALLENGER.

this opinion may be advanced that—at any rate as far as I know—hitherto no specimen of one of the known species of *Colossendeis* has been caught with egg-masses on its ovigerous legs. Considering that they are not the eggs of the *Colossendeis* itself, it becomes almost impossible to form an opinion as to the animal they belong to. Among the gastropodous molluscs numerous forms are known, which construct egg-capsules, and attach them to foreign bodies. Perhaps the present capsules belong to an animal of that group. That the long legs of our animals may easily be mistaken by other animals for dead bodies is shown, I believe, by the fact that numerous other animals, which cannot be considered as parasites, and which, as a rule, are found on stones, shells of molluscs, carapaces of crabs, &c., fix themselves on these legs. So a small sponge and a poly-zoon are on *Nymphon brachyrhynchus*, a stalk-like process most probably of a tubularian polyp is found on the leg of a *Colossendeis*; a species of *Scalpellum* is extremely numerous on the legs of *Nymphon robustum*, Bell. Of the numerous specimens of this species collected in Barents Sea, which I have investigated, there is not a single one with these ectoparasites. But on the other hand, they are very common on the hundreds of specimens of this species which were obtained by the "Knight-Errant." Professor G. O. Sars enumerates in his two latest papers on the Crustaceans of the Norwegian Expeditions numerous species of *Scalpellum*, found at higher northern latitudes, but he does not mention that they are found on the legs of the most common Pycnogonid of the North Atlantic and North Polar Sea. Moreover, a preliminary comparison of this species of *Scalpellum* shows differences with those described. I therefore believe it to be a new one, and wish to name it *Scalpellum nymphocola*.

SUMMARY OF THE REPORT.

1. Of the forty-one species of Pycnogonida dredged during the voyage of H.M.S. Challenger and the cruise of the "Knight-Errant" thirty-three are new to science.

2. Of the nine genera represented in those collections three are new.

3. Those genera which range most widely geographically are also those which range most widely in depth.

4. There are deep-sea species, but true deep-sea genera do not seem to exist.

5. The Pycnogonida form a distinct and very natural group (class) of arthropodous animals. Their common progenitor (their typical form) must be considered as a hypothetical Pycnogonid with three-jointed mandibles, multi-jointed palpi, and ovigerous legs with numerous rows of denticulate spines on the last joints.

6. This class of the Arthropoda may be thus characterised :—Arthropoda breathing by the general surface of the body, which body consists of a cephalothoracic, three thoracic segments, and a rudimentary abdominal segment. The cephalic part of the cephalothoracic segment bears anteriorly a proboscis, consisting of three coalesced parts, one præ-oral (labrum ?), two post-oral ones (mandibles ?), and three pair of cephalic appendages, the first two of which in the adult state sometimes have become rudimentary, the third pair being always present at least in one of the two sexes. The first pair of these appendages represents the antennæ, the two others are post-oral. The thoracic part of the cephalothoracic segment and the three thoracic segments are each furnished with a pair of long eight-jointed legs, into which the alimentary canal sends off long cœca.

7. The function of the integumentary cavities is primarily respiratory.

8. The typical form of the nervous system shows a supracœsophageal and five thoracic ganglia. The supracœsophageal ganglion gives off the nerves for the mandibles (antennæ), the integumentary nerves, and a strong nerve for the proboscis. Besides these it probably gives off nerves for the intestine (sympathic nerves). The first thoracic ganglion consists of two coalesced ganglia, and gives off four pairs of nerves, two pairs innervating the proboscis, then the pair of palpar nerves, and finally those for the ovigerous legs. The following four ganglia give off the nerves for the four pairs of legs ; the last ganglion gives sometimes two sometimes one pair of nerves for the abdomen.

9. In addition to the nerves mentioned above the proboscis is innervated by three strong bundles of nerves and ganglia united by a stronger and some feebler secondary œsophageal nerve rings.

10. In some genera the inner surface of the integument is covered by a net-work of nerves and ganglia in connection with, and most probably issuing from, the integumentary nerves given off by the supraœsophageal ganglion.

11. In the most primitive condition the eye of the Pycnogonid consists of a rounded transparent part of the integument, the inner surface of which is furnished with some small ganglia and nerve-fibres issuing from the integumentary nerve bundle. The highly developed eye of the shallow-water species shows ganglionic cells, distinct retinal rods, and a lens consisting of a thickened part of the chitinous skin of the animal.

12. Those eyes which have lost their pigment and their retinal rods are rudimentary. They cannot be considered as forming the transition between the highly-developed eye and its most primitive condition.

13. That part of the œsophagus which runs through the proboscis has the function of a masticating apparatus. Where the œsophagus enters the intestinal tract (the stomach) small glands (pancreatic, most probably) are present.

14. The original condition of the genital glands is in the form of a U-shaped mass, placed above the intestine and giving off branches which penetrate the legs. Whereas for the male glands the original form prevails in most (all ?) genera, for the female glands it seems to be a rule that only the lateral parts entering the legs are developed. The genital pores of the females are larger than those of the males ; they are found ventrally towards the extremity of the second joint of the leg. Whereas for the females it is the rule that these pores are present on all the legs ; it often happens in the males that they are only present on the two or three hindmost pair of legs.

15. There are always distinct vasa efferentia, but there are not always true oviducts.

16. In *Nymphon brevicaudatum*, Miers, females also bear the eggs on the ovigerous legs.

17. The larva creeping out of the egg is already furnished with an azygous outgrowth of the region surrounding the mouth (the proboscis). As a rule in that stage only three pairs of appendages (the later cephalic ones) are present.

18. These larvæ are often furnished on their mandibles with an apparatus producing a single or numerous threads, wherewith the young is attached to the ovigerous leg of its parent.

19. About the relation in which the Pycnogonida stand to either the Crustacea or the Arachnida we know as much or as little as we do about the relation in which these two classes Arthropoda stand to each other.

Note.—While I was engaged in preparing the index of this report, and after the rest of it had been printed off, Mr Edmund B. Wilson of Baltimore kindly sent me two papers which he had recently published. In one (the Pycnogonida of New England and Adjacent Waters, Report of the United States Commissioner of Fish and Fisheries, part vi. for 1878, pp. 463–506, pls. i.–vii.) the author gives an account of the present knowledge of the species of Pycnogonida known to occur on the coasts of New England and Nova Scotia. With two exceptions (*Achelia scabra*, Wilson, and *Nymphon macrum*, Wilson) the species here described are the same as those of a former paper by Mr Wilson, published in the Trans. Connect. Acad. Sci., vol. v. pp. 1–26, 1880. The new *Achelia* is quite unknown to me, but *Nymphon macrum*, Wils., is undoubtedly the species which I have described in my report (p. 45) as *Nymphon brevicollum*. The Challenger specimens were taken south of Halifax (83 fathoms), those described by Mr Wilson in the Gulf of Maine (85 to 115 fathoms).

The other paper (Reports on the Results of Dredging, under the Supervision of Alexander Agassiz, along the East Coast of the United States, during the summer of 1880, by the United States' Coast Survey Steamer "Blake," Commander J. R. Bartlett, U.S.N., commanding. xiii. Report on the Pycnogonida, by Edmund B. Wilson; Bulletin of the Museum of Comparative Zoölogy at Harvard College, vol. viii., No. 12, Cambridge, Mass., March 1881, pp. 239–256, pls. i.–v.) contains descriptions of ten species of Pycnogonids, five of which are new. These belong to three genera, two of which are considered by the author as new. The new species are in the first place two species of *Colossendeis*, Jarzynsky, *Colossendeis colossea*, and *Colossendeis macerrima*. Then a new genus *Scaeorhynchus*, with the species *Scaeorhynchus armatus*, is proposed; finally, the new genus *Pallenopsis*, with the species *Pallenopsis forficifer* and *Pallenopsis longirostris*, is described. The descriptions are illustrated by very good figures. On comparing these figures and descriptions with those of my report, there can be little doubt that *Colossendeis colossea* and *C. macerrima* are very nearly related to, if not identical with, my *Colossendeis gigas* and *C. leptorhynchus*. As to the genus *Scaeorhynchus*, I do not think there are sufficient grounds for separating it from *Ascorhynchus*, G. O. Sars. Neither the presence of dactyli on the first pair of legs, nor the structure of the rudimentary mandibles (antennae) makes it proper to separate these genera: *Scaeorhynchus* (like *Gnamptorhynchus*, Böhm) is only a synonym of *Ascorhynchus*. The species *armatus*, Wilson, seems to be different from those hitherto described, and also from those of the present report.

The new genus *Pallenopsis* is intended to embrace those species which come near to *Phoxichilidium*, but which are characterised by ten-jointed accessory legs present in both sexes, and by three-jointed mandibles. Three (perhaps four) species described in my report show these characters also, and (pp. 82 and 88) I have been long in doubt whether I should not propose a new genus for these species. I did not take the step because I do not wish to augment the number of genera more than necessary until our knowledge of generical characters is more perfect. Mr Wilson is not so slow in proposing new genera; in the present instance, I believe, however, that his proposal has a fair chance of being accepted. The two species described by Mr Wilson are, I believe, different from those described in my report.

I regret very much that in my report species will be found mentioned, described, and figured as new, which at the date of its publication will have been already described. As I was aware of the large collections of deep-sea animals collected by Professor Alexander Agassiz, and felt sure that my report, with its numerous plates, would take a considerable time in passing through the press, I took the liberty of writing to Professor Agassiz, to ask him to whom the working-out of the Pycnogonids of his latest cruises had been entrusted. He kindly complied with my request, and informed me that the Pycnogonids along with the Crustacea had been sent to Professor Alphonse Milne-Edwards in Paris. I then addressed Professor Milne-Edwards, sending him at the same time proof-copies of the plates of my report on which the new species were figured, and as he favoured me with an answer, in which he promised to make use of the names proposed by me, I had every reason to believe that zoological literature, at least in the case of the deep-sea Pycnogonida, would not be encumbered by synonyms. Where I have not been successful in this respect I hope nobody will lay the blame upon me.

LEIDEN, 10th May 1881.

EXPLANATION OF PLATES.

PLATE I.

Nymphon hamatum, n. sp. (figs. 1-9).

Fig. 1. Male, dorsal view; magnified 3 diameters.
 „ 2. Male, ventral view; magnified 7 diameters.
 „ 3. The claws of the mandibles; magnified 41 diameters.
 „ 4. The last four joints of the ovigerous legs; magnified 34 diameters.
 „ 5. The denticulate spines of the ovigerous legs; magnified 272 diameters.
 „ 6. The second coxal joint of the leg of a female; magnified 6 diameters.
 „ 7. The tubercles on the thigh of the male; magnified 94 diameters.
 „ 8. The hook-like process at the end of the thigh; magnified 41 diameters.
 „ 9. The last two joints and the claw of one of the legs; magnified 21 diameters.

PLATE II.

Nymphon longicoxa, n. sp. (figs. 1-5).

Fig. 1. Male, dorsal view; magnified 7¾ diameters.
 „ 2. The last four joints of the palpus; magnified 20 diameters.
 „ 3. The claws of the mandibles; magnified 94 diameters.
 „ 4. The denticulate spines of the ovigerous legs; magnified 272 diameters.
 „ 5. The last two joints and the claw of one of the legs; magnified 41 diameters.

Nymphon compactum, n. sp. (figs. 6-8).

Fig. 6. Female, ventral view; magnified 6½ diameters.
 „ 7. Part of the body of a female, dorsal view; magnified 8 diameters.
 „ 8. The denticulate spines of the ovigerous legs; magnified 272 diameters.

Nymphon procerum, n. sp. (figs. 9-12).

Fig. 9. Female, dorsal view; magnified 6¾ diameters.
 „ 10. The claws of the mandibles; magnified 94 diameters.
 „ 11. The claw of the ovigerous leg; magnified 94 diameters.
 „ 12. The denticulate spines of the ovigerous legs; magnified 272 diameters.

PLATE III.

Nymphon longicollum, n. sp. (figs. 1-3).

Fig. 1. Male, dorsal view ; magnified 6¾ diameters.

,,   2. The same, seen ventrally ; magnified 20 diameters.

,,   3. The claws of the mandibles ; magnified 47 diameters.

Nymphon meridionale, n. sp. (figs. 4-8).

Fig. 4. Male, dorsal view ; magnified 20 diameters.

,,   5. The claws of the mandibles ; magnified 94 diameters.

,,   6. Claw and last part of the tenth joint of the ovigerous leg; magnified 272 diameters.

,,   7. The last two joints of the leg ; magnified 41 diameters.

,,   8. The claw with one of the secondary claws of one of the legs ; magnified 136 diameters.

Nymphon grossipes, Fabr. (sp.), (figs. 9-12).

Called *Nymphon armatum*, n. sp., at the foot of the plate.

Fig. 9. Male, ventral view ; magnified 6¾ diameters.

,,   10. The claws of the mandibles ; magnified 41 diameters.

,,   11. The denticulate spines of the ovigerous legs ; magnified 575 diameters.

,   12. The last joint of the leg of a female ; magnified 41 diameters.

Nymphon brevicollum, n. sp. (figs. 13-15).

Fig. 13. Female, ventral view ; magnified 6¾ diameters.

.,   14. Claw and last part of the tenth joint of the ovigerous leg; magnified 272 diameters.

,,   15. Claw with one of the secondary claws of the leg ; magnified 94 diameters.

PLATE IV.

Nymphon grossipes, Fabr. (sp.) (fig. 1).

Called *Nymphon armatum*, n. sp., at the foot of the plate.

Fig. 1. Oculiferous tubercle ; magnified 41 diameters.

Nymphon brachyrhynchus, n. sp. (figs. 2-7).

Fig. 2. Male, dorsal view ; magnified 41 diameters.

,,   3. Female, ventral view ; magnified 6¾ diameters.

Fig. 4. The claws of the mandibles; magnified 94 diameters.
„ 5. Palpus; magnified 41 diameters.
„ 6. Claw of the ovigerous leg; magnified 272 diameters.
„ 7. Claw of one of the legs; magnified 94 diameters.

Nymphon fuscum, n. sp. (figs. 8–11).

Fig. 8. Male, ventral view; magnified 6¾ diameters.
„ 9. The claws of the mandibles; magnified 94 diameters.
„ 10. Palpus; magnified 41 diameters.
„ 11. Claw of the ovigerous leg; magnified 180 diameters.

Nymphon brevicaudatum, Miers (figs. 12–13).
Called *Nymphon hispidum*, n. sp., at the foot of the plate.

Fig. 12. Male, dorsal view; magnified 12 diameters.
„ 13. Male, ventral view; magnified 6 diameters.

PLATE V.

Nymphon brevicaudatum, Miers (figs. 1–5).
Called *Nymphon hispidum*, n. sp., at the foot of the plate.

Fig. 1. Palpus; magnified 41 diameters.
„ 2. Claws of the mandibles; magnified 41 diameters.
„ 3. Last five joints of the ovigerous leg; magnified 41 diameters.
„ 4. Denticulate spine of the ovigerous leg; magnified 272 diameters
„ 5. Last two joints of the leg; magnified 34 diameters.

Nymphon perlucidum, n. sp. (figs. 6–10).

Fig. 6. The front part of the body, seen ventrally; magnified 41 diameters.
„ 7. The whole animal, dorsal view; magnified 6¾ diameters.
„ 8. The last four joints of the ovigerous legs; magnified 94 diameters.
„ 9. Two of the denticulate spines of the ovigerous leg; magnified 272 diameters.
„ 10. The last two joints of the leg; magnified 36 diameters.

Ascorhynchus orthorhynchus, n. sp. (figs. 11–13).

Fig. 11. Dorsal view; magnified 3½ diameters.
„ 12. Ventral view; magnified 2 diameters.
„ 13. Palpus; magnified 21 diameters.

PLATE VI.

Ascorhynchus orthorhynchus, n. sp. (figs. 1–4).

Fig. 1. Rudimentary mandible ; magnified 94 diameters.
,, 2. The last four joints of the ovigerous leg ; magnified 21 diameters.
,, 3. The denticulate spines of one of the joints of the ovigerous leg ; magnified 272 diameters.
,, 4. The last two joints of the leg ; magnified 21 diameters.

Ascorhynchus glaber, n. sp. (figs. 5–9).

Fig. 5. Dorsal view ; magnified 5¼ diameters.
,, 6. Mandibles ; magnified 21 diameters.
,, 7. The mandible of a young specimen ; magnified 94 diameters.
,, 8. The last four joints of the ovigerous leg ; magnified 21 diameters.
,, 9. The last joint of the leg ; magnified 21 diameters.

Ascorhynchus minutus, n. sp. (figs. 10–16).

Fig. 10. Ventral view ; magnified 7¼ diameters.
,, 11. Side view ; magnified 7¾ diameters.
,, 12. The last three joints of the ovigerous leg ; magnified 70 diameters.
,, 13. The eighth joint of the ovigerous leg ; magnified 182 diameters.
,, 14. The last two joints of the leg ; magnified 70 diameters.
,, 15. The last joint of the first leg ; magnified 94 diameters.
,, 16. Hairs of the palpus ; magnified 272 diameters.

PLATE VII.

Oorhynchus aucklandiæ, n. sp. (figs. 1–7).

Fig. 1. Ventral view ; magnified 15 diameters.
,, 2. Dorsal view ; magnified 7½ diameters.
,, 3. The front part seen dorsally ; magnified 34 diameters.
,, 4. Palpus ; magnified 94 diameters.
,, 5. The last four joints of the ovigerous leg ; magnified 130 diameters.
,, 6. One of the legs ; magnified 34 diameters.
,, 7. The last two joints of the leg ; magnified 70 diameters.

Discoarachne brevipes, n. sp. (figs. 8–12).

Fig. 8. Dorsal view; magnified 8 diameters.
,,   9. Ventral view; magnified 8 diameters.
,, 10. Palpus; magnified 94 diameters.
,, 11. The last four joints of the ovigerous leg; magnified 94 diameters.
,, 12. The last two joints of the leg; magnified 94 diameters.

PLATE VIII.

Colossendeis gigas, n. sp. (figs. 1–2).

Fig. 1. Ventral view; natural size.
,, 2. Dorsal view of the body; natural size.

Colossendeis leptorhynchus, n. sp. (figs. 3–7).

Fig. 3. Ventral view; natural size.
,, 4. Dorsal view of the body; natural size.
,, 5. The last four joints of the palpus; magnified 30 diameters.
,, 6. The last four joints of the ovigerous legs; magnified 21 diameters.
,, 7. The claw of one of the legs; magnified 48 diameters.

PLATE IX.

Colossendeis megalonyx, n. sp. (figs. 1–3).

Fig. 1. Lateral view; magnified 2 diameters.
,, 2. The last three joints of the palpus; magnified 39 diameters.
,, 3. The arrangement of the denticulate spines on the ninth joint of the ovigerous leg; magnified 94 diameters.

Colossendeis robusta, n. sp. (figs. 4, 5).

Fig. 4. Dorsal view; magnified 2 diameters.
,, 5. The arrangement of the denticulate spines on the ninth joint of the ovigerous leg; magnified 41 diameters.

Colossendeis gracilis, n. sp. (figs. 6–8).

Fig. 6. Ventral view; magnified 5 diameters.
,, 7. The last four joints of the palpus; magnified 41 diameters.
,, 8. The arrangement of the denticulate spines on the ninth joint of the ovigerous leg; magnified 235 diameters.

THE VOYAGE OF H.M.S. CHALLENGER.

PLATE X.

Colossendeis gigas, n. sp. (figs. 1–5).

Fig. 1. The arrangement of the denticulate spines at the ninth joint of the ovigerous leg; magnified 41 diameters.
„ 2. One of the denticulate spines of the ninth joint of the ovigerous leg of an old specimen; magnified 136 diameters.
„ 3. The same of an outside row of a younger specimen; magnified 136 diameters.
„ 4. The same of an inside row of a younger specimen; magnified 136 diameters.
„ 5. Two rudimentary spines of a row quite to the inside of an old specimen; magnified 136 diameters.

Colossendeis gracilis, n. sp. (figs. 6–7).

Fig. 6. Specimen with mandibles, dorsal view; magnified 6½ diameters.
„ 7. The arrangement of the denticulate spines on the ninth joint of the ovigerous leg; magnified 94 diameters.

Colossendeis brevipes, n. sp. (figs. 8–9).

Fig. 8. The last five joints of the palpus; magnified 20 diameters.
„ 9. The arrangement of the denticulate spines on the ninth joint of the ovigerous leg; magnified 94 diameters.

Colossendeis minuta, n. sp. (figs. 12–14).

Fig. 12. Dorsal view; magnified 6 diameters.
„ 13. The last five joints of the palpus; magnified 41 diameters.
„ 14. The arrangement of the denticulate spines on the ninth joint of the ovigerous leg; magnified 272 diameters.

PLATE XI.

Pallene australiensis, n. sp. (figs. 1–7).

Fig. 1. Ventral view; magnified 7 diameters.
„ 2. Dorsal view; magnified 17 diameters.
„ 3. Claws of the mandibles; magnified 94 diameters.
„ 4. Last five joints of the ovigerous leg of the male; magnified 34 diameters.
„ 5. Ninth joint of the ovigerous leg of the male; magnified 270 diameters.
„ 6. Last two joints of the leg; magnified 65 diameters.
„ 7. Spine on the sixth joint of the leg; magnified 270 diameters.

Pallene lœvis, n. sp. (figs. 8-12).

Fig. 8. Female specimen, dorsal view ; magnified 6 diameters.
„ 9. Female specimen, ventral view ; magnified 17 diameters.
„ 10. Last four joints of the ovigerous leg ; magnified 65 diameters.
„ 11. Last joint of the ovigerous leg ; magnified 272 diameters.
„ 12. Last two joints of the leg ; magnified 41 diameters.

PLATE XII.

Pallene languida, n. sp. (figs. 1-5).

Fig. 1. Ventral view ; magnified 41 diameters.
„ 2. Dorsal view ; magnified 41 diameters.
„ 3. Last joint of the mandible ; magnified 94 diameters.
„ 4. Last five joints of the ovigerous leg ; magnified 94 diameters.
„ 5. Denticulate spines of the ovigerous leg ; magnified 575 diameters.

Phoxichilidium patagonicum, n. sp. (figs. 6-9).

Fig. 6. Dorsal view, natural size.
„ 7. Ventral view ; magnified 6 diameters.
„ 8. Last four joints of the ovigerous leg ; magnified 36 diameters.
„ 9. Last two joints of the leg ; magnified 6 diameters.

Phoxichilidium patagonicum, var. *elegans*, Hoek (fig. 10).

Fig. 10. Ventral view ; magnified 8 diameters.

PLATE XIII.

Phoxichilidium oscitans, n. sp. (figs. 1-5).

Fig. 1. Dorsal view ; magnified 4 diameters.
„ 2. Front part of the body, seen ventrally ; magnified 8 diameters.
„ 3. The mouth seen from the front ; magnified 7 diameters.
„ 4. Last four joints of the ovigerous leg ; magnified 41 diameters.
„ 5. Last two joints of the leg ; magnified 8 diameters.

Phoxichilidium mollissimum, n. sp. (figs. 6-9).

Fig. 6. Lateral view ; magnified 4 diameters.
„ 7. Mouth, front view ; magnified 9 diameters.

Fig.   8. Last four joints of the ovigerous leg ; magnified 28 diameters.
  „   9. Part of the sixth joint of the leg ; magnified 8 diameters.

Phoxichilidium pilosum, n. sp. (figs. 10–13).

Fig. 10. Dorsal view ; magnified 5 diameters.
  „   11. Front part of the body, ventral view ; magnified 9 diameters.
  „   12. Last four joints of the ovigerous leg ; magnified 41 diameters.
  „   13. Last two joints of the leg ; magnified 9 diameters.

PLATE XIV.

Phoxichilidium fluminense, Kröyer (figs. 1–4).

Fig.   1. Ventral view ; magnified 8 diameters.
  „   2. Oculiferous tubercle, dorsal view ; magnified 28 diameters.
  „   3. Last five joints of the ovigerous leg ; magnified 56 diameters.
  „   4. Last two joints of the leg ; magnified 28 diameters.

Phoxichilidium insigne, n. sp. (figs. 5–7).

Fig.   5. Dorsal view ; magnified 8½ diameters.
  „   6. Front part of the body, seen ventrally ; magnified 28 diameters.
  „   7. Last two joints of the leg ; magnified 56 diameters.

Hannonia typica, n. gen., n. sp. (figs. 8–11).

Fig.   8. Dorsal view ; magnified 17 diameters.
  „   9. Ventral view ; magnified 5½ diameters.
  „   10. Last four joints of the ovigerous leg ; magnified 70 diameters.
  „   11. Last two joints of the leg ; magnified 30 diameters.

PLATE XV.

Nymphon macronyx, G. O. Sars. (figs. 1–7).

Fig.   1. Ventral view ; magnified 7 diameters.
  „   2. Oculiferous tubercle ; magnified 64 diameters.
  „   3. Mandible ; magnified 41 diameters.
  „   4. Palpus ; magnified 41 diameters.
  „   5. Denticulate spines of the ovigerous legs ; magnified 272 diameters.
  „   6. Claw of the ovigerous leg ; magnified 272 diameters.
  „   7. Last two joints of the leg ; magnified 41 diameters.

Nymphon longicoxa, Hoek (figs. 8, 9).

Fig. 8. Spines at the sixth joint of the ovigerous leg of a male; magnified 272 diameters.

„ 9. Genital opening of the male; magnified 136 diameters.

Nymphon compactum, Hoek (fig. 10).

Fig. 10. Last two joints of the leg; magnified 7 diameters.

Nymphon longicollum, Hoek (fig. 11).

Fig. 11. Auxiliary claws; magnified 272 diameters.

Nymphon brevicollum, Hoek (figs. 12, 13).

Fig. 12. Articulation (?) in the fifth joint of the ovigerous leg of the male; magnified 94 diameters.

„ 13. The ovigerous leg of the male; magnified 7 diameters.

Ascorhynchus orthorhynchus, Hoek (figs. 14, 15).

Fig. 14. Mouth, front view; magnified 5 diameters.

„ 15. Genital pore of the female; magnified 64 diameters.

Ascorhynchus glaber, Hoek (fig. 16).

Fig. 16. Articulation of the proboscis; magnified 6 diameters.

PLATE XVI.

Structure of the integument of *Nymphon robustum*, Bell (figs. 1–3).

Fig. 1. Transverse section of the integument; magnified 272 diameters. *a*, spine; *b*, respiratory cavity; *c*, setæ; *d*, nerve; *e*, blood-corpuscles; *f*, epithelium.

„ 2. Setæ; magnified 575 diameters.

„ 3. Setæ; magnified 575 diameters.

Structure of the integument of *Nymphonhamatum*, Hoek (figs. 4–6).

Fig. 4. Surface of the fourth joint of the leg of a male; magnified 94 diameters. *g*, pores.

„ 5. Structure of the gland; magnified 272 diameters.

„ 6. Transverse section of the thigh of a male; magnified 94 diameters. *g*, pore; *k*, blood; *i*, intestinal cœcum; *h*, gland.

Structure of the integument of *Nymphon brevicaudatum*, Miers (fig. 7).

Fig. 7. Transverse section of the fourth joint of the leg of a female; magnified 56 diameters. *i*, as in fig. 6; *l*, mature egg; *m*, immature eggs; *x*, mud.

Structure of the integument of *Ascorhynchus glaber*, Hoek (figs. 8–10).

Fig. 8. Transverse section of a part of the integument ; magnified 272 diameters. *b, c, e*, as in fig. 1.

„   9. Transverse section of the thigh of a male ; magnified 56 diameters. *n*, testis ; *g, h, i, k*, as in fig 6.

„ 10. Cells from the glandular mass of the male ; magnified 272 diameters.

Genital organs, &c., of *Ascorhynchus orthorhynchus*, Hoek (fig. 11).

Fig. 11. Transverse section of the thigh of a female ; magnified 56 diameters. *k, m, h, i*, as in figs. 6 and 7.

Structure of the integument of *Colossendeis leptorhynchus*, Hoek (figs. 12, 13).

Fig. 12. Transverse section of a part of the integument ; magnified 170 diameters. *b, e*, as in fig. 1.

„ 13. Spine ; magnified 170 diameters.

Structure of the integument of *Colossendeis megalonyx*, Hoek (fig. 14).

Fig. 14. Transverse section through the thigh of a male ; magnified 41 diameters. *h, i*, as in fig. 6.

Structure of the integument, &c., of *Colossendeis leptorhynchus*, Hoek (figs. 15, 16).

Fig. 15. Transverse section through the thigh of a male ; magnified 272 diameters. *o*, vesicle ; *p*, wounded canal.

„ 16. Transverse section of the thigh of a female ; magnified 56 diameters. *m, i*, as in fig. 7.

Structure of the integument of *Phoxichilidium patagonicum*, Hoek (fig. 17).

Fig. 17. Transverse section of the integument of a female ; magnified 170 diameters. *b, c, f*, as in fig. 1.

Structure of the integument, &c., of *Phoxichilidium insigne*, Hoek (fig. 18).

Fig. 18. Transverse section of the thigh of a male ; magnified 94 diameters. *h, i, k*, as in fig. 6.

PLATE XVII.

Fig. 1. *Colossendeis proboscidea*, Sab. (sp.); dorsal half of the body ; magnified 6 diameters. *a*, intestine ; a^1, a^2, two short cœca not reaching beyond the cephalothorax ; *b*, muscles moving the proboscis ; *c*, heart ; *d*, anus.

Fig. 2. *Colossendeis proboscidea*, Sab. (sp.) ; ventral half of the body with the nervous system ; magnified 6 diameters. *a*, parts of the intestine ; *p*, palpus ; I. II.- V., first to fifth thoracic ganglia ; *n.p.* and *n.pr.*, two branches of the palpar nerve ; *n.o.l.*, nerve for the ovigerous leg ; n^1–n^4, four nerves for the legs ; *n.a.*, nerves for the abdomen.

„ 3. *Nymphon robustum*, Bell ; ventral half with the nervous system ; magnified 6 diameters. *a*, parts of the intestine, cœca, &c. ; *m.u.*, muscles running from the one segment to the other ; *n.p.*, palpar nerve ; *n.pr.*, proboscideal nerve ; *n.o.l.*, n^1–n^4, *n.a.*, as in fig. 2.

„ 4. *Nymphon robustum*, Bell ; supracesophageal ganglion with its nerves ; magnified 6 diameters. *a*, intestine ; *m*, mandibles ; *g.s.o.*, supracesophageal ganglion ; *c*, œsophageal commissures ; *n.m.*, nerve for the mandibles ; *u.p.n.*, unpair proboscideal nerve ; *p.p.n.*, small proboscideal nerves ; *g.p.*, proboscideal ganglion.

„ 5. *Nymphon robustum*, Bell ; dorsal half of the body showing the intestine ; magnified 4 diameters. *a*, intestine ; *e*, œsophagus.

„ 6. *Nymphon robustum*, Bell ; part of the intestine ; magnified 47 diameters. *a*, intestine ; *e*, œsophagus ; *gl*, intestine glands.

PLATE XVIII.

Fig. 1. Integumentary cavities of *Colossendeis leptorhynchus*, Hoek ; ♀ specimen ; first tibial joint ; magnified 287 diameters. *i*, part of the wall of the intestinal cœcum ; *c*, septa of connective tissue ; *b*, blood-corpuscles ; *d*, and *d'*, glandular cells.

„ 2. Transverse section of the fourth joint of the leg of *Colossendeis proboscidea*, Sab. (sp.) ; ♂ specimen ; magnified 47 diameters. *i*, intestinal cœcum ; *g*, glands ; *t*, testis ; *c*, fibres of connective tissue.

„ 3. Part of the integument of *Colossendeis proboscidea*, Sab. (sp.) ; fourth joint of the leg ; ♂ specimen ; magnified 272 diameters. *c*, integumentary cavity ; *g*, gland ; *d*, duct of the gland.

„ 4. Œsophageal ring of *Colossendeis megalonyx*, Hoek ; magnified 40 diameters. *s.* supracesophageal ganglion ; *c*, commissures ; *t*, first thoracic ganglion ; *pr.* azygous proboscideal nerve ; *m*, mandibular nerve ; *o*, optic nerves (?) ; *t'*, first nerve arising from the first thoracic ganglion ; *t''*, second nerve (one of the three main proboscideal nerves), dividing into two branches after penetrating the proboscis ; *pa.*, palpar nerve (dividing in two branches close to the ganglion) ; *o*, nerve for the ovigerous leg ; *c'*, commissures between the first and second thoracic ganglia.

Fig.　5. Innervation of the rudimentary eyes of *Nymphon robustum*, Bell; magnified 94 diameters. *n*, optic nerves forming the ganglion *g*; *t*, oculiferous tubercle ; *o*, rudimentary eyes.

„　6. Innervation of the proboscis of *Nymphon robustum*, Bell; magnified 10 diameters. The proboscis has been figured as transparent; the mandible and the palp of the left side have been cut away. *t*, oculiferous tubercle ; *s*, supraœsophagal ganglion ; *c*, œsophagal commissures ; *u*, azygous proboscideal nerve ; *p*, secondary proboscideal nerves arising from the supraœsophageal ganglion ; *mn*, mandibular nerves ; *p*, proboscideal nerves arising from the first thoracic ganglion ; *g*, ganglionic bundles ; *x*, small ganglia in the front of the proboscis ; *y*, large proboscideal ganglia ; *a*, first proboscideal nerve ring ; a^{I}–a^{IV} second to fifth proboscideal ring.

„　7. Transverse section of cephalothoracic segment of *Colossendeis leptorhynchus*, Hoek (female specimen) ; magnified 10 diameters. *s*, supraœsophageal ganglion ; *t*, first thoracic ganglion ; *œ*, œsophagus ; *c*, œsophageal commissures ; *i*, rudimentary cœca of the intestine ; *m*, and *m'*, muscle-bundles ; *m*, passing through the œsophageal commissures.

„　8. Part of the ganglionic bundle in the proboscis of *Nymphon robustum*, Bell ; magnified 94 diameters. *u*, *g*, *a*, *a'*, *a''*, *x*, and *y*, as in fig. 6.

„　9. Transverse section of the proboscis of *Nymphon robustum*, Bell ; magnified 23 diameters. I., II., and III. the three chitinous plates limiting the œsophageal cavity ; *t*, transverse muscles ; *l*, longitudinal muscle-bundles ; *a*, proboscideal ring ; *g*, ganglion of the ganglionic bundle ; *p*, the three main proboscideal nerves on transverse section.

„　10. One of the ganglia on the inner surface of the integument of *Colossendeis proboscidea*, Sab. (sp.) ; magnified 272 diameters. *c*, integumentary cavities ; *b*, blood-corpuscles.

„　11. Six sections through the supraœsophageal and first thoracic ganglia of *Nymphon strömii*, Kröyer ; magnified 56 diameters. *s*, supraœsophageal ganglion ; *œ*, œsophagus ; *t*, first thoracic ganglion ; *i*, intestinal cœcum for the mandibles ; *c*, fibres of connective tissue ; *o*, optic lobes ; *o'*, part of the optic nerves ; *m*, mandibular lobes ; *x*, azygous lobe of unknown function ; *co*, œsophageal commissures ; *p*, origin of the two main proboscideal nerves ; *l*, origin of the nerves for the ovigerous legs ; *co'*, commissures between the first and second thoracic ganglia.

„　12. Longitudinal section through the first thoracic ganglia of *Nymphon brachyrhynchus*, Hoek ; magnified 170 diameters. *pr*, proboscideal nerves ; *pa*, part of the neurilemma of the palpar nerve ; *c*, part of the neurilemma of the ovigerous nerve ; *c*, commissure between this and the second thoracic ganglion.

PLATE XIX.

Fig. 1. Egg-mass of *Nymphon brevicollum*, Hoek; magnified 16 diameters.

,, 2. Transverse section of egg-mass of *Nymphon brevicollum*, Hoek; magnified 41 diameters. *o*, opening occupied by the ovigerous leg; *s*, small pores left between the eggs; *c*, the substance soldering the eggs together; *m*, mud.

,, 3. Transverse section of the egg of *Nymphon brevicollum*, Hoek; magnified 272 diameters.

,, 4. Structure of the protoplasm of one of the segments of the same egg; magnified 575 diameters.

,, 5. Cells of the blastoderm and segments of the nutritive yolk of *Nymphon brevicaudatum*, Miers; magnified 272 diameters.

,, 6. Nuclei of the cells of the blastoderm of the egg of *Nymphon brevicaudatum*, Miers; magnified 575 diameters.

,, 7. Transverse section through the egg of *Nymphon brevicaudatum*, Miers; magnified 94 diameters.

,, 8. Embryo of *Nymphon brachyrhynchus*, Hoek, showing the three pairs of appendages and the proboscis; magnified 94 diameters.

,, 9. Front part of the embryo of *Nymphon brevicaudatum*, Miers, seen ventrally; magnified 135 diameters.

,, 10. The embryo of *Nymphon brevicaudatum*, Miers, ventral view; magnified 94 diameters.

,, 11. Transverse section through the egg of *Nymphon brevicaudatum*, Miers; magnified 94 diameters. *a*, mandible; *b*, first pair of legs; *c, d*, second and third pair of cephalic appendages.

,, 12. Larva of *Nymphon brevicollum*, Hoek, dorsal view; magnified 94 diameters.

,, 13. Larva of *Nymphon brevicollum*, Hoek, ventral view; magnified 94 diameters.

PLATE XX.

Fig. 1. Larva of *Nymphon brevicollum*, Hoek; magnified 94 diameters.

,, 2. Spinning apparatus in the mandible of the larva of *Nymphon brevicollum*, Hoek; magnified 94 diameters.

,, 3. Front part of the larva of *Nymphon hamatum*, Hoek; magnified 94 diameters.

,, 4. Spinning apparatus in the mandible of the larva of *Nymphon hamatum*, Hoek; magnified 272 diameters.

,, 5. Spinning apparatus in the mandible of *Nymphon longicoxa*, Hoek; magnified 272 diameters.

,, 6. Larva of *Ascorhynchus minutus*, Hoek; magnified 272 diameters.

(ZOOL. CHALL. EXP.—PART X.—1881.) K 21

4

PLATE XXI.

Fig. 1. Eye of *Nymphon strömii*, Kröyer, transverse section ; magnified 272 diameters.

„ 2. Eye of *Nymphon brachyrhynchus*, Hoek, longitudinal section, taken laterally ; magnified 170 diameters.

„ 3. Eye of *Nymphon brachyrhynchus*, Hoek, longitudinal section, passing nearly through the middle ; magnified 170 diameters.

„ 4. Rods and ganglion cells from the eye of *Nymphon brachyrhynchus*, Hoek ; magnified 575 diameters.

„ 5. Rods (isolated) from the eye of *Nymphon brachyrhynchus*, Hoek ; magnified 575 diameters.

„ 6. Œsophagus of *Nymphon hamatum*, Hoek, transverse section ; magnified 272 diameters.

„ 7. Glands at the end of the œsophagus of *Nymphon hamatum*, Hoek ; magnified 94 diameters. *a*, the glands ; *s*, supraœsophageal ganglion ; *f*, first thoracic ganglion.

„ 8. Glands at the end of the œsophagus of *Nymphon hamatum*, Hoek, transverse section ; magnified 170 diameters. *a*, the glands ; *b*, the wall of the intestine.

„ 9. Group of fat cells from *Nymphon robustum*, Bell ; magnified 272 diameters.

„ 10. Male genital organs of *Colossendeis proboscidea*, Sab. (sp.), natural size. *a*, testis ; *b*, intestine.

„ 11. The opening of the male genital organs in the second joint of the leg of *Colossendeis proboscidea*, Sab. (sp.). *a*, testis ; *b*, intestine ; *c*, vas efferens.

„ 12. Transverse section of the testis of *Colossendeis proboscidea*, Sab. (sp.) ; magnified 94 diameters.

„ 13. Histological structure of the wall of the vas efferens of *Colossendeis proboscidea*, Sab. (sp.) ; magnified 575 diameters.

„ 14. Transverse section of the body of *Colossendeis leptorhynchus*, Hoek, about between the second and third leg ; magnified 20 diameters. *h*, heart ; i^1-i^3, intestine ; *o*, ovary ; *n*, nervous system.

„ 15. Ovarian egg of *Nymphon brevicaudatum*, Miers ; magnified 94 diameters.

„ 16. Section of the ovary in the fourth joint of the leg of *Nymphon robustum*, Bell ; magnified 94 diameters. *i*, wall of the intestinal cœcum ; *c*, cavity completely filled with eggs, when the animal is mature.

„ 17. Egg of *Nymphon longicoxa*, Hoek ; magnified 94 diameters.

„ 18. Section through the fourth joint of the leg of *Phoxichilidium patagonicum*, Hoek ; magnified 41 diameters.

INDEX.

Abdomen, 16.
Acarus, 2.
Accessory claws, 16.
Achelia, 3, 9, 23, 24, 26, 27, 61.
    echinata, 26.
    hispida, 27.
    lævis, 27.
    scabra, 147.
    spinosa, 24, 26.
Alcinous, 5, 9, 26.
    vulgaris, 26.
    megacephalus, 26.
Affinities of the Pycnogonids with the other Arthropoda, 108.
Alimentary canal and its appendages, 124.
Ammothea, 9, 23, 24.
    achelioides, 24.
    brevipes, 24.
    carolinensis, 23.
    longipes, 23.
    pycnogonoides, 4, 23.
Anatomy of the Pycnogonida, 100.
Anomorhynchus smithii, 99.
Anoplodactylus, 5, 82.
    lentus, 33.
Antliata, 2.
Apertures of the heart, number of the, 127.
Appendages, development of the, 138.
    food-yolk in the, 138.
Arachnida, 2.
Aranea, 2.
Ascorhynchus, 4, 9, 10, 23, 25, 53, 147.
    abyssi, 4, 25, 55.
    glaber, 8, 25, 37, 53, 55.
    minutus, 7, 25, 55, 58.
    orthorhynchus, 7, 25, 57.
    ramipes, 25, 56, 58.

Bdella, 2.
Blastoderm, 137, 138.
Bell, Th., referred to, 3.
Blood-corpuscles, 128.
Blood-corpuscles in the integumentary cavities, 103.
Body of Pycnogonid, description of, 14.
    (ZOOL. CHALL. EXP.—PART X.—1881.)

Böhm, R., referred to, 4, 6, 49, 51, 82, 134.
Böhmia, 9, 24.
    chelata, 24.
"Borstenapparat" of the integument, 102.
Bottom, nature of the, 12.
Brachiopoda, 8.
Brünnich, 2.
Buchholz, R., referred to, 3.

Cavanna, G., referred to, 5, 122, 128, 134.
Cavities in the integument, 101.
Cephalothoracic segment, 14.
Circulatory apparatus, 127.
Claparède, A. R. E., referred to, 4.
Claws, at the ends of the legs, 16.
Cleavage of the egg, 136.
Commensals of the legs of Pycnogonids, 144.
Cœca of the intestine, number and structure of, 8, 126.
COLOSSENDEIDÆ, 23.
Colossendeis, 3, 6, 9, 10, 23, 28, 59, 61, 70, 98.
    angusta, 4, 28, 65.
    borealis, 28, 98.
    brevipes, 8, 28, 29, 70, 72, 73.
    colossea, 147.
    gigas, 8, 28, 37, 61, 64, 65, 66, 67, 147.
    gigas-leptorhynchus, 28, 65.
    gracilis, 8, 28, 29, 37, 59, 69, 71, 72, 73, 74.
    kröyerii, 28.
    leptorhynchus, 7, 28, 37, 64, 66, 67, 74, 147.
    macerrima, 147.
    media, 8, 28, 29, 70, 71, 73.
    megalonyx, 7, 11, 28, 67.
    minuta, 8, 11, 29, 73.
    proboscidea, 7, 11, 28, 69, 98.
    robusta, 7, 28, 66.
Commissures uniting the thoracic ganglia, 116.
Commissures, circum-œsophageal, 109.
Copulation observed, 131.
Corniger lilyenborfi, 27.
Costa, O. G., referred to, 5.

K 22

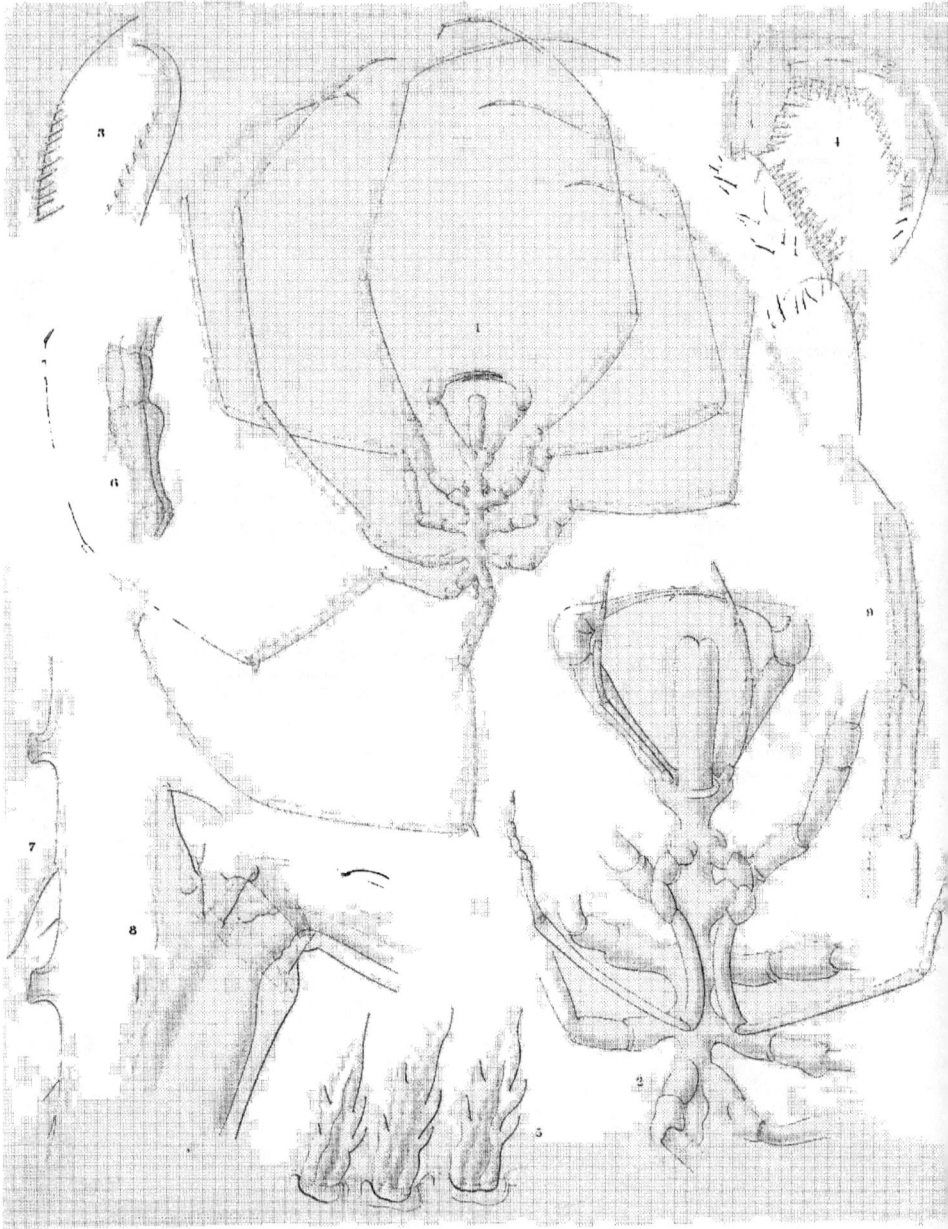

NYMPHON HAMATUM n. sp.

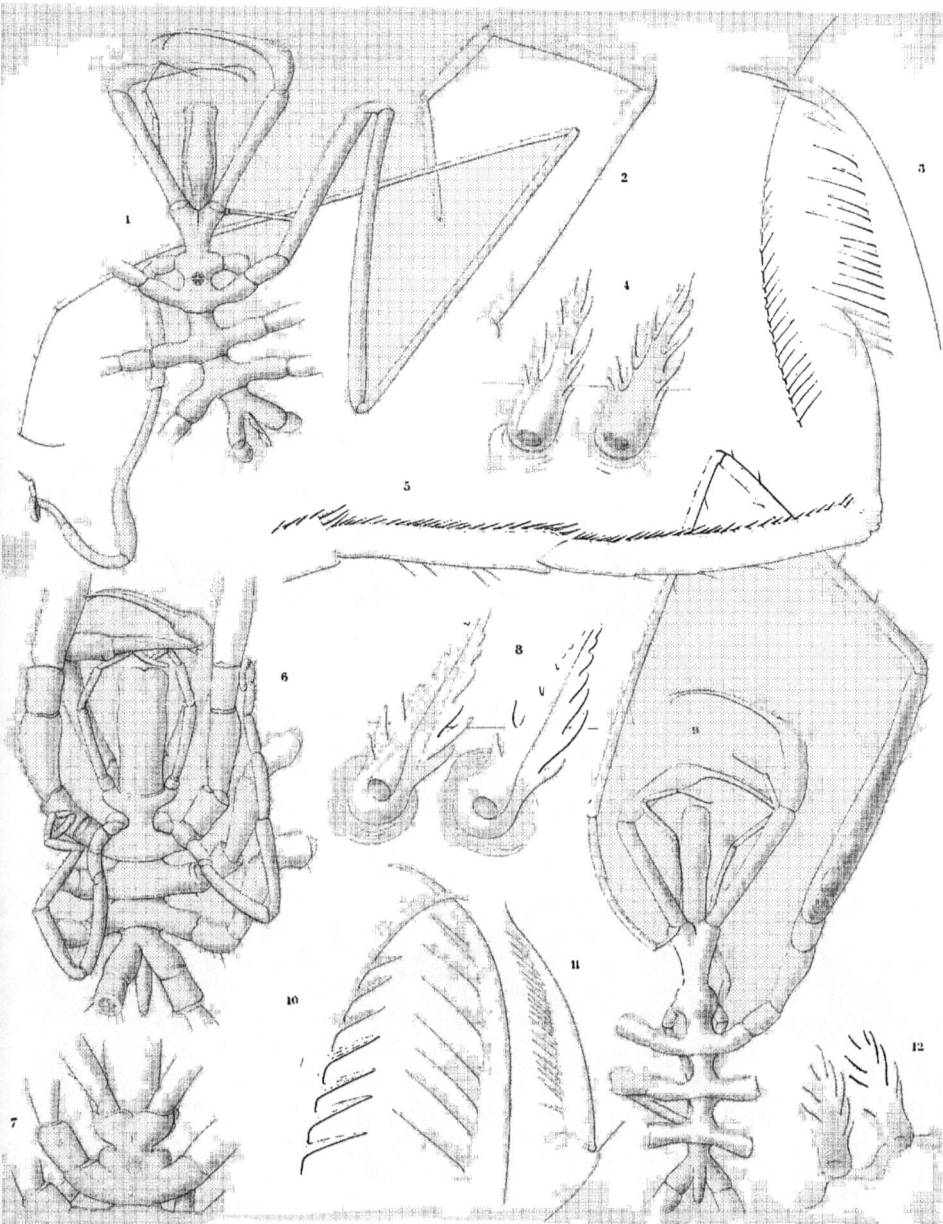

1-5 NYMPHON LONGICOXA n sp; 6-8 N. COMPACTUM n sp. 9-12 N. PROCERUM n sp.

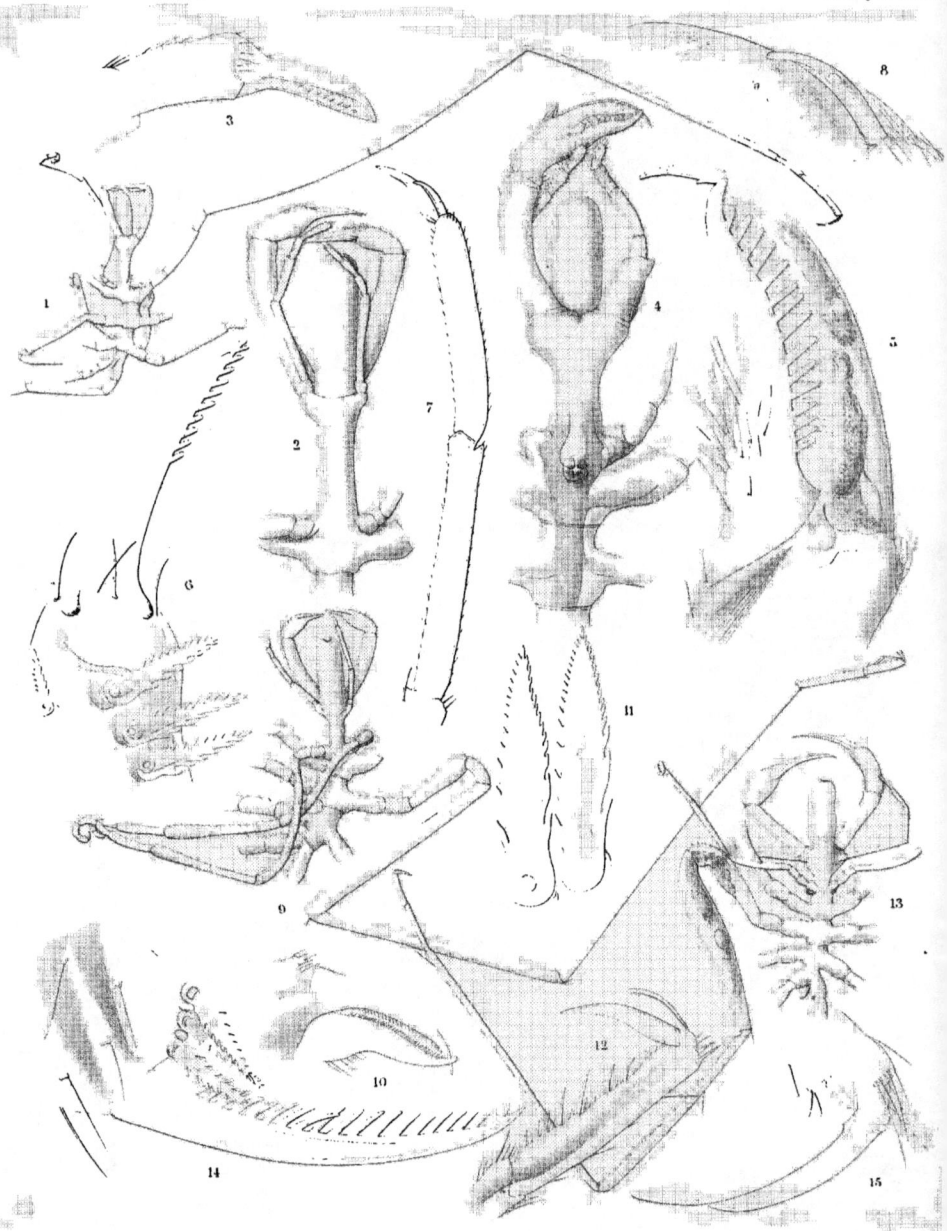

1-3 NYMPHON LONGICOLLUM n sp 4-8 N. MERIDIONALE n sp 9-12 N. ARMATUM n sp 13-15 N. BREVICOLLUM n sp

I NYMPHON ARMATUM n sp 2-7 N BRACHYRHYNCHUS n sp 8-11 N FUSCUM n sp 12-13 N HISPIDUM n sp.

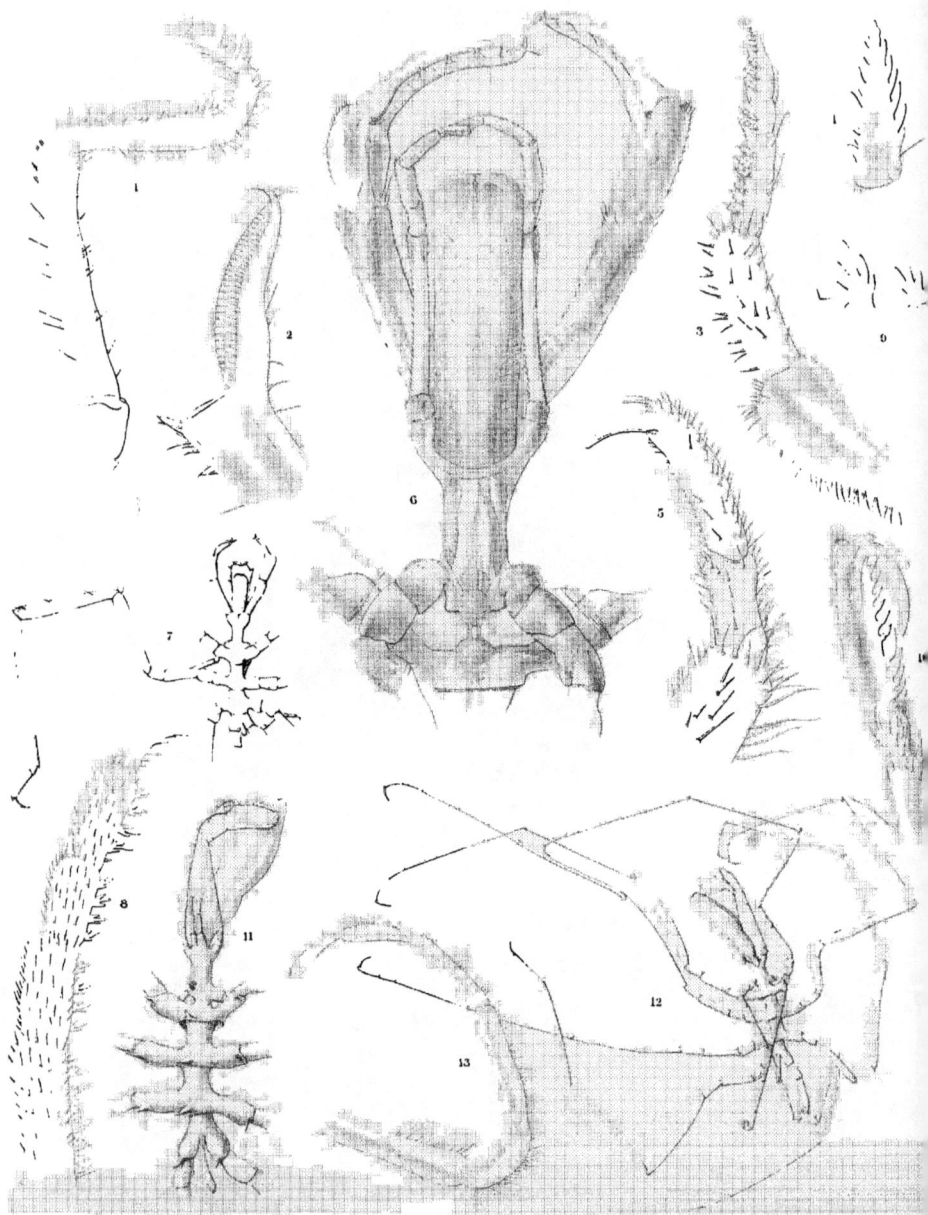

C. Hava Del.   Ad Wendel. Lith

1-5 NYMPHON HISPIDUM n.sp, 6-10 N. PERLLCIDUM n.sp 11-13 ASCORHYNCHUS ORTHORHYNCHUS n.sp

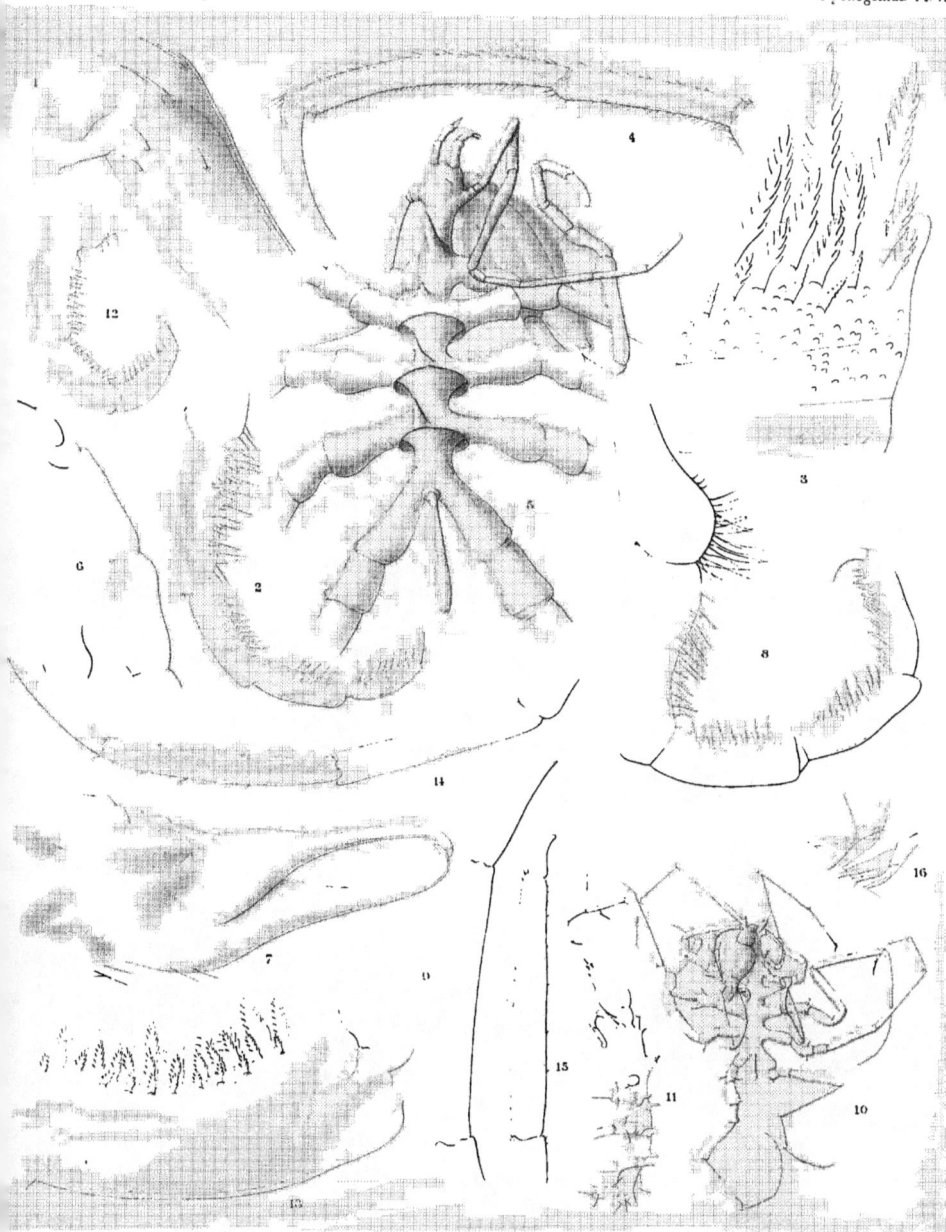

1-4 ASCORHYNCHUS ORTHORHYNCHUS n sp 5-9 A GLABER n. sp. 10-16 A. MINUTUS n sp.

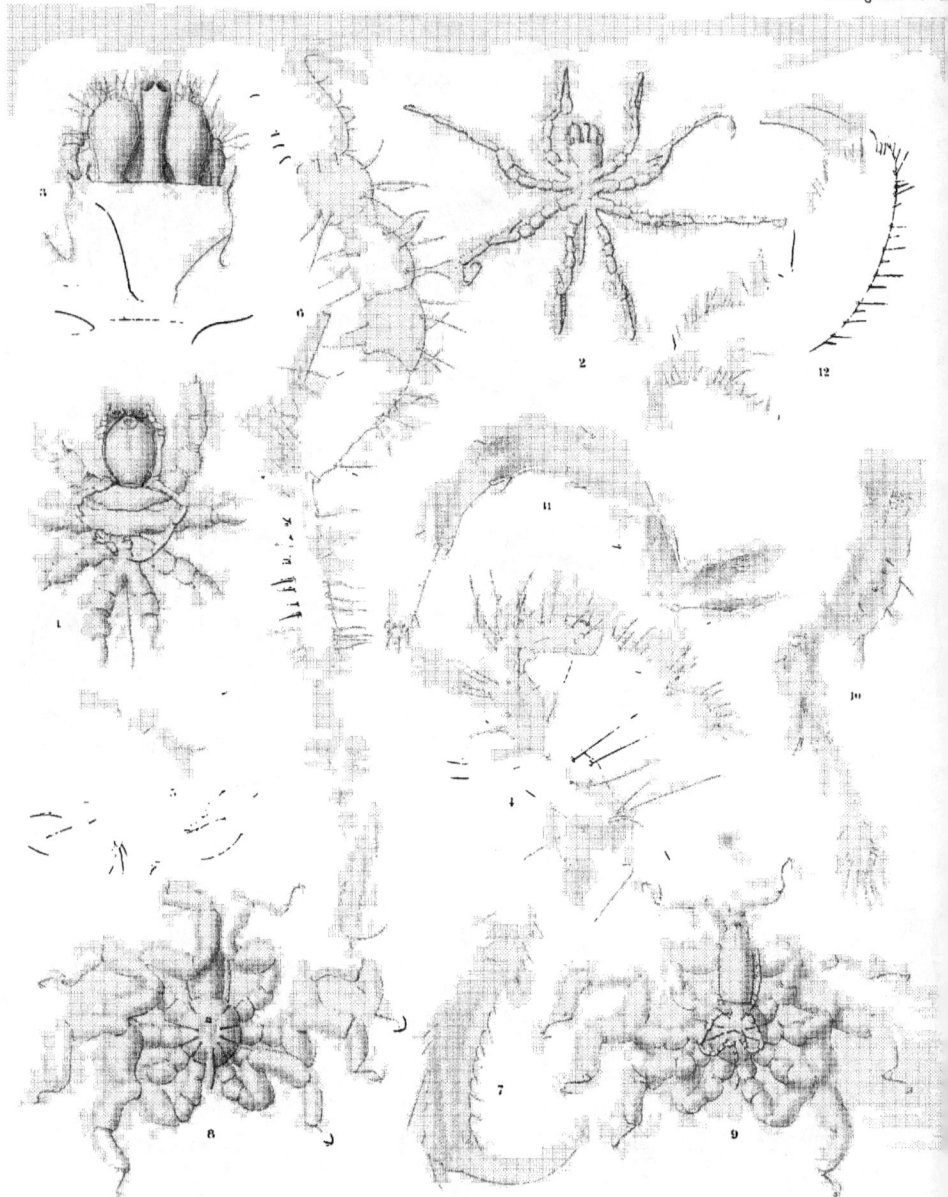

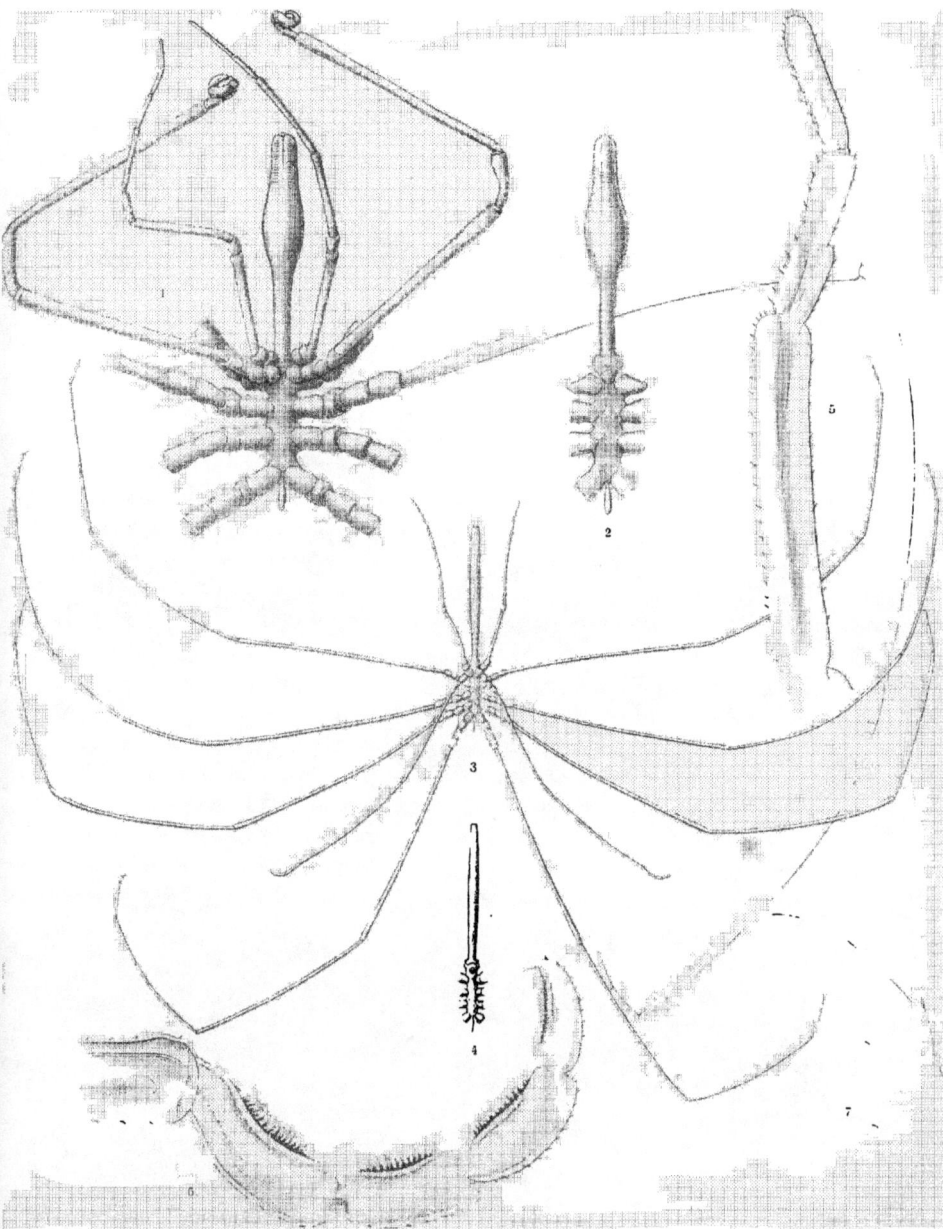

1-2 COLOSSENDEIS GIGAS n sp. 4-7 C. LEPTORHYNCHUS n sp

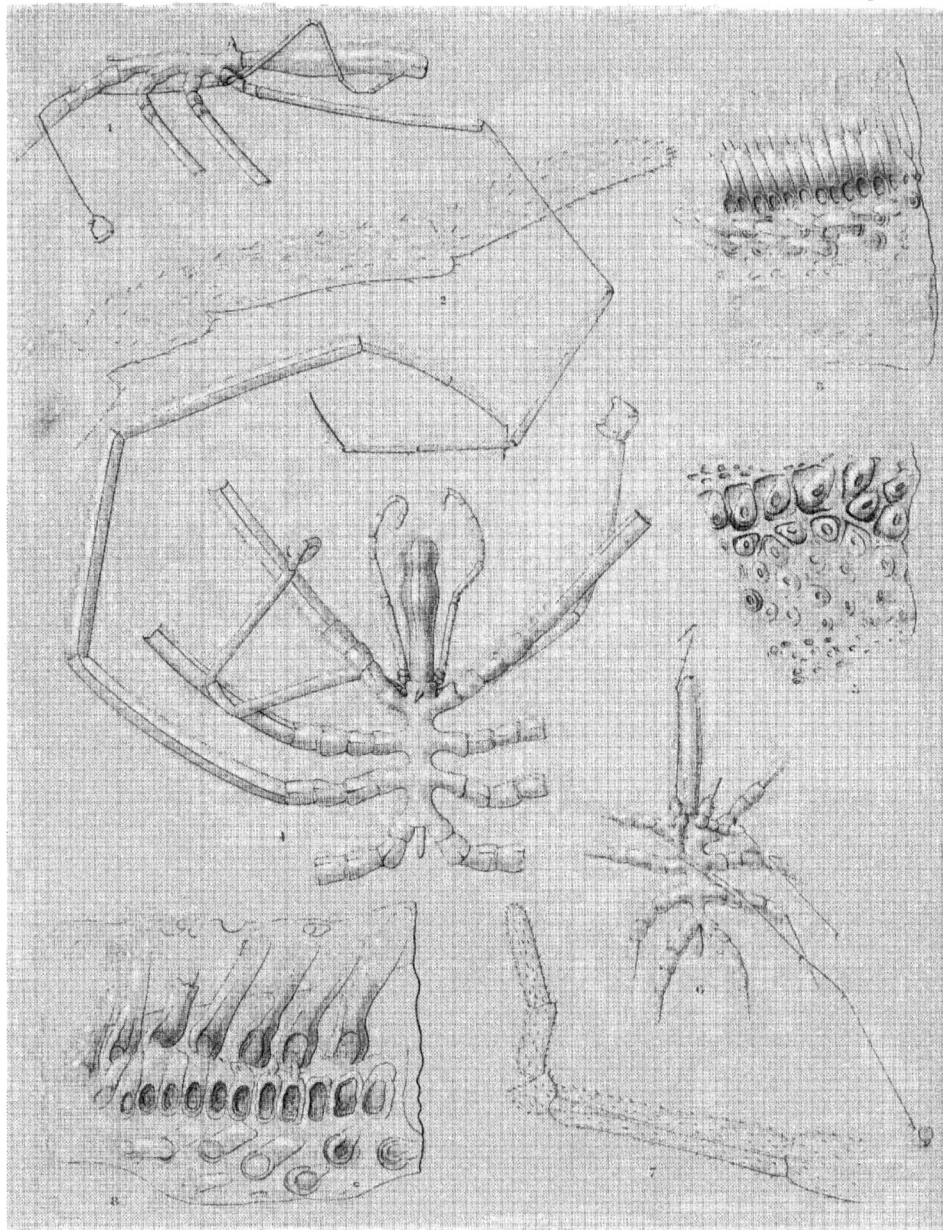

1-3 COLOSSENDEIS MEGALONYX n sp. 4-5 C ROBUSTA n sp 6-8 C. GRACILIS n sp

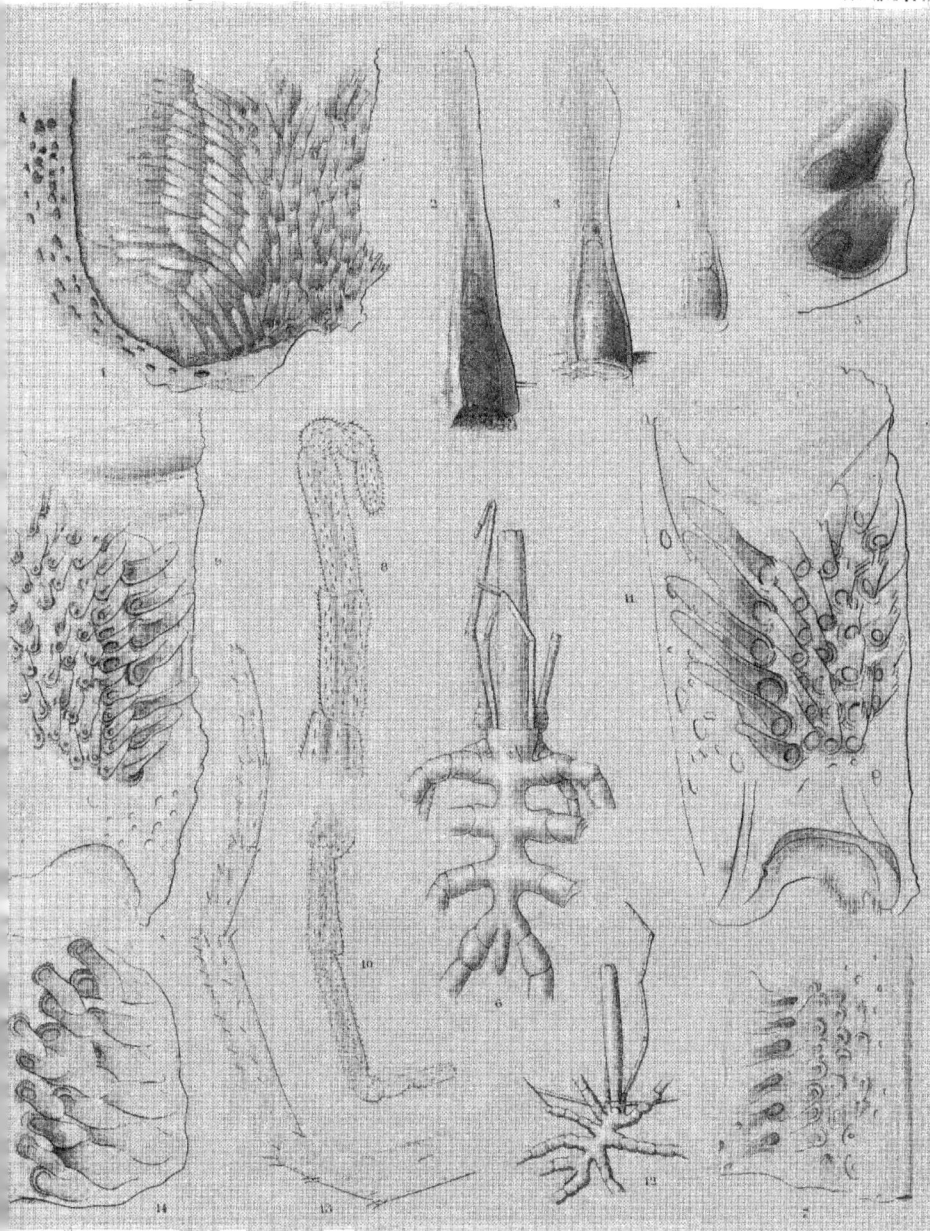

1-5 COLOSSENDEIS GIGAS n. sp. 6-7 C. GRACILIS n. sp 8-9 C. BREVIPES n. sp.
10-11 C. MEDIA n. sp. 12-14 C. MINUTA n. sp

1-7 PALLENE AUSTRALIENSIS n. sp. 8-12 P. LAEVIS n sp

1-5 PALLENE LANGUIDA n. sp. 6-9 PHOXICHILIDIUM PATAGONICUM n sp 10 PH PATAGONICUM var elegans

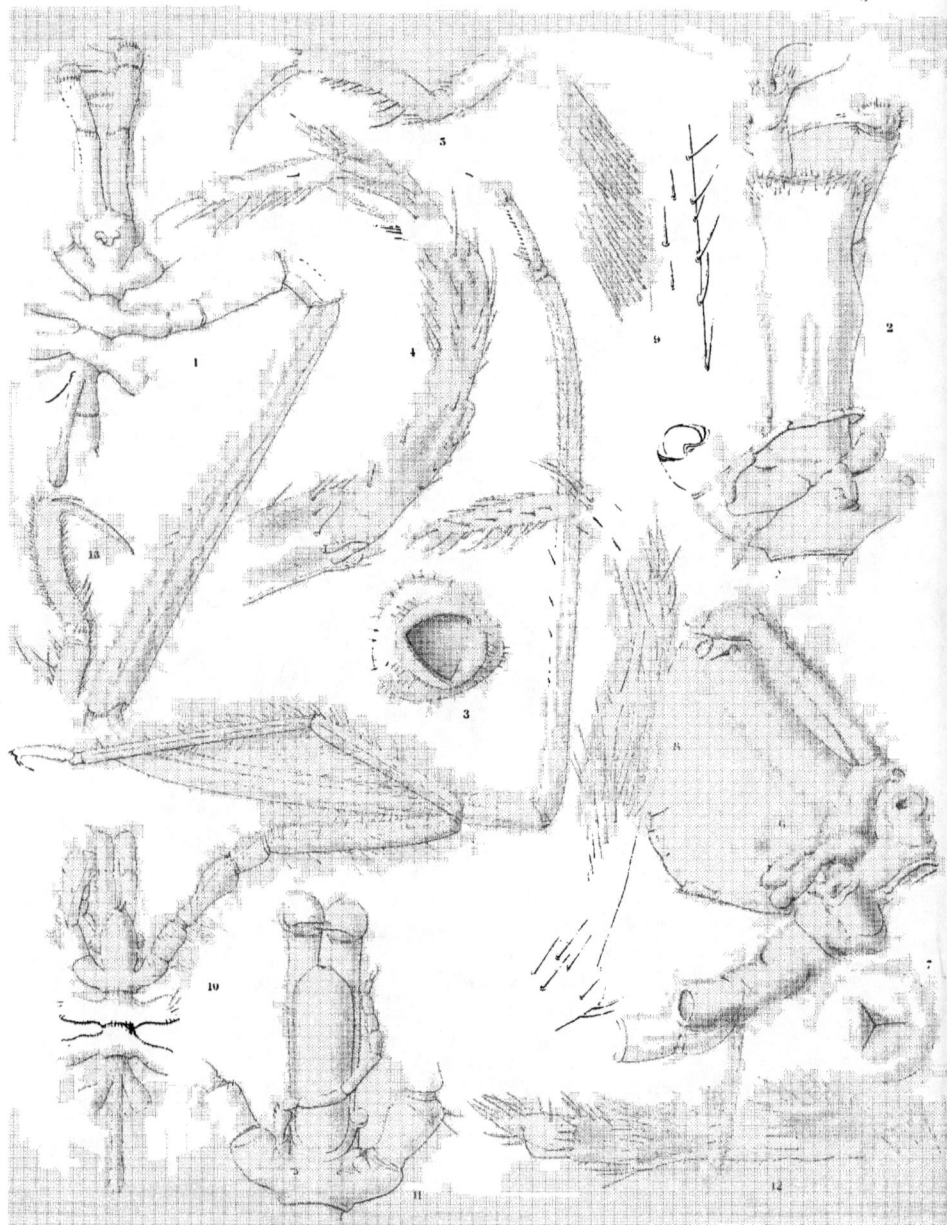

1-5 PHOXICHILIDIUM OSCITANS n sp. 6-9 PH. MOLLISSIMUM n sp 10-13 PH. PILOSUM n sp

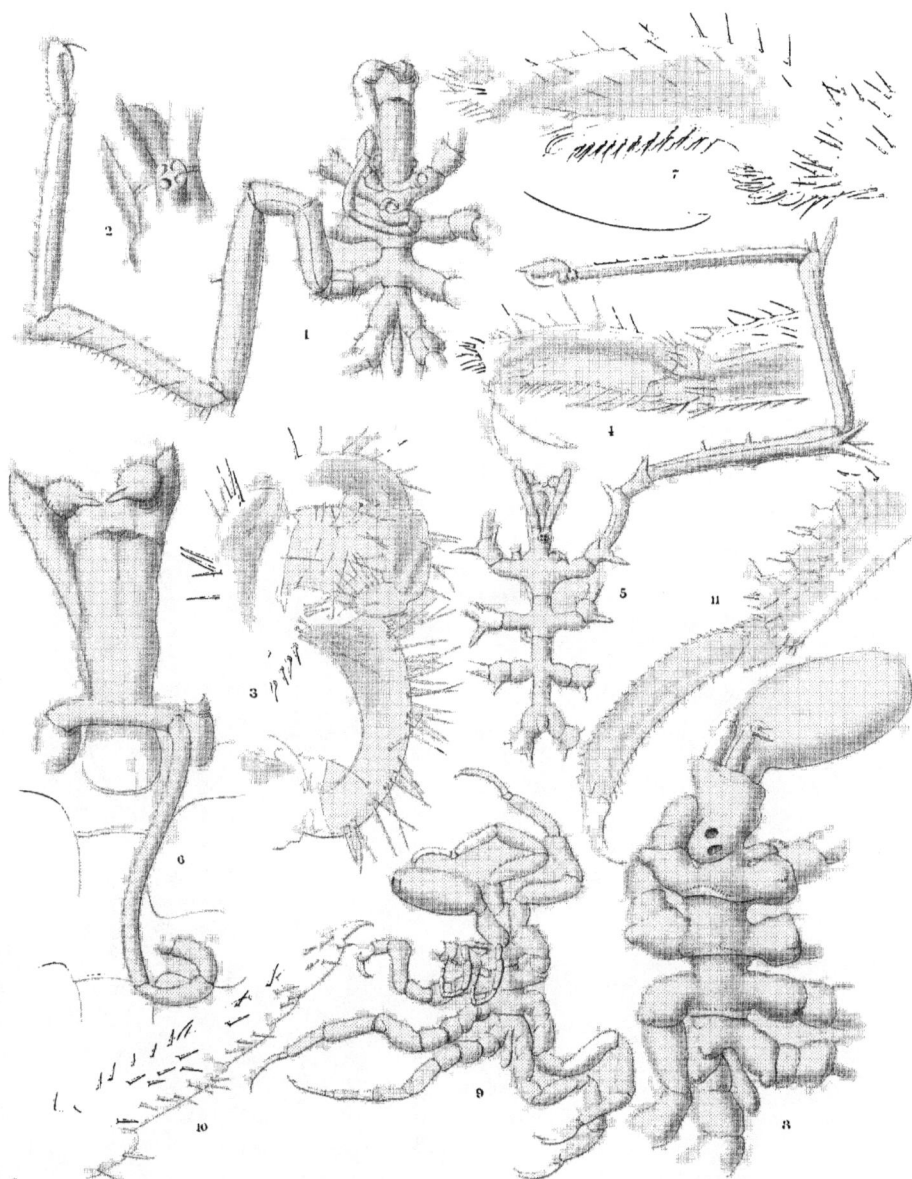

1-4 PHOXICHILIDIUM FLUMINENSE Kröyer. 5-7 PH INSIGNE n sp 8-11 HANNONIA TYPICA n. gen n sp

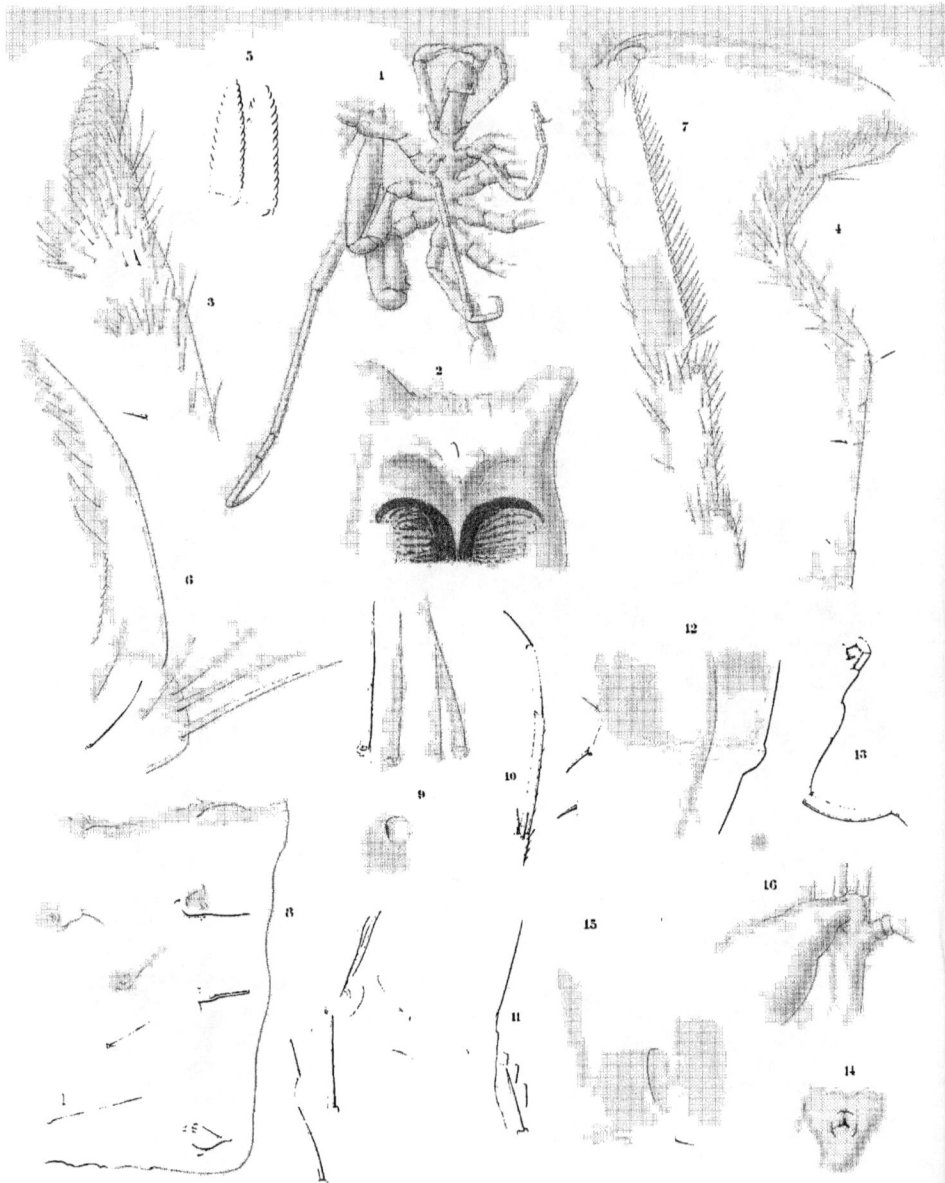

1-7 NYMPHON. 8-11 ASCORHYNCHUS. 12-16 COLOSSENDEIS. 17-18 PHOXICHILIDIUM.

1-2 COLOSSENDEIS. 3-6 NYMPHON.

1-4. 7. 10 COLOSSENDEIS 5-6. 8-9. 11-12 NYMPHON

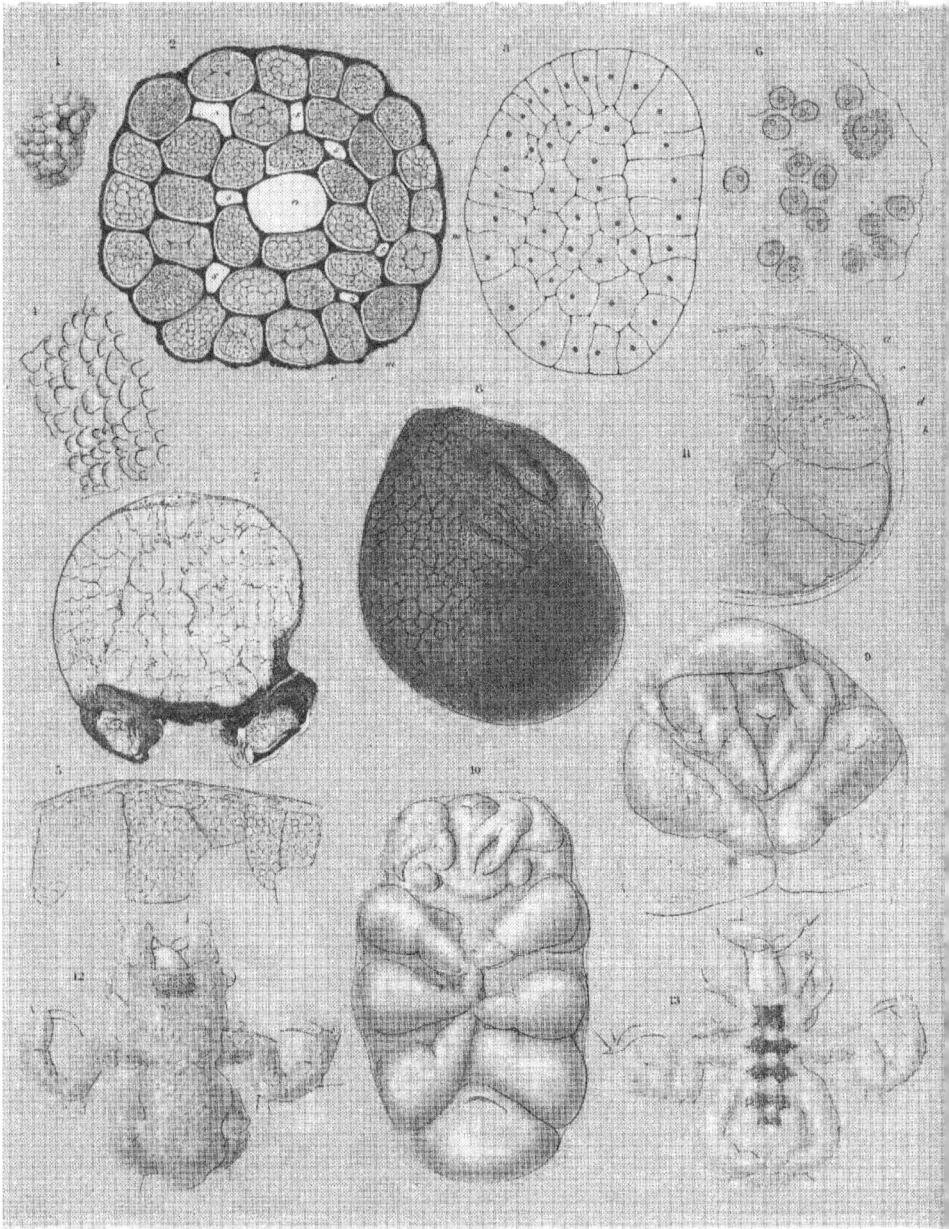

EMBRYOLOGY OF NYMPHON

EMBRYOLOGY OF NYMPHON AND ASCORHYNCHUS.

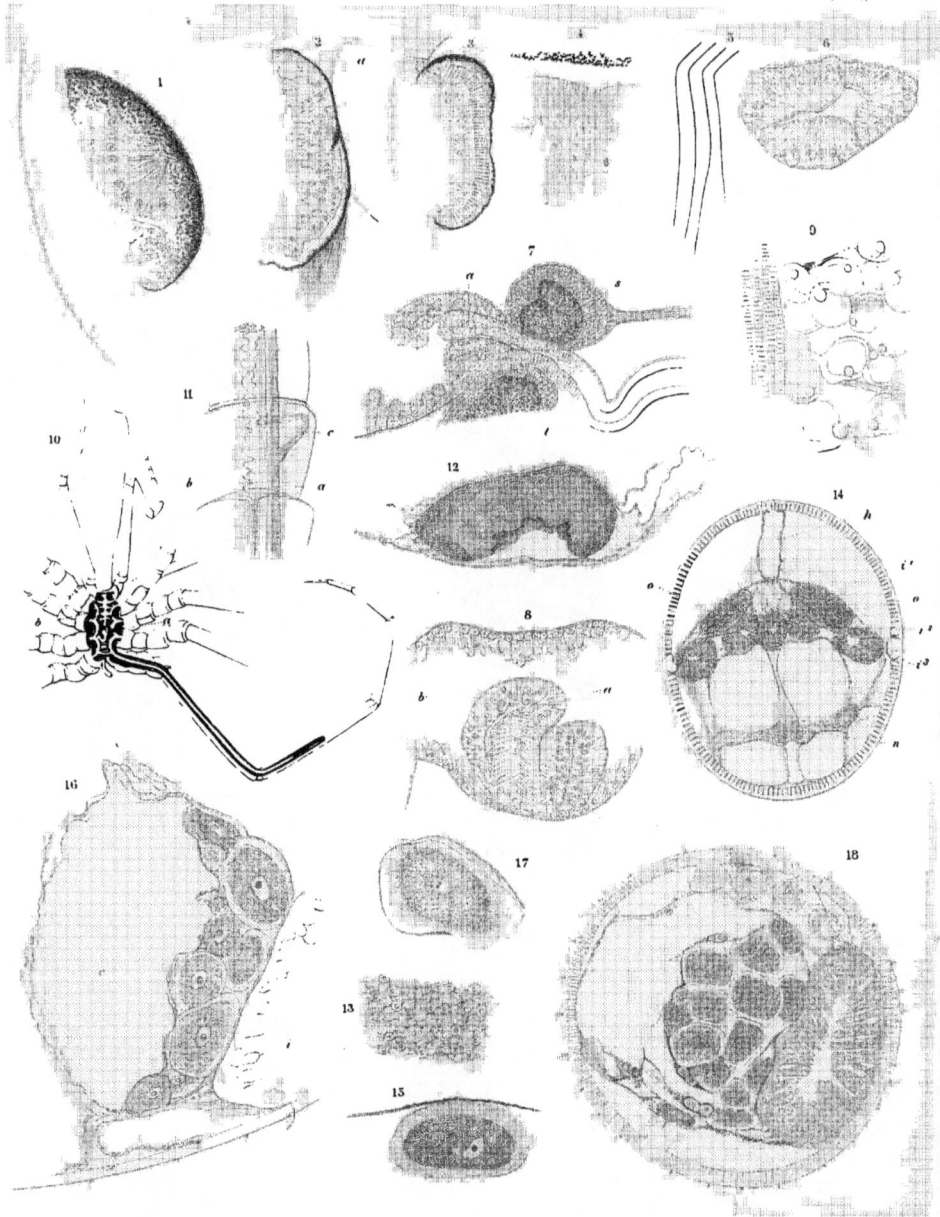

1-9, 15-17 NYMPHON. 10-14 COLOSSENDEIS. 18 PHOXICHILIDIUM.

www.ingramcontent.com/pod-product-compliance
Lightning Source LLC
Chambersburg PA
CBHW021801190326
41518CB00007B/405

* 9 7 8 3 7 4 1 1 8 4 4 2 0 *